SpringerBriefs in Applied Sciences and Technology

W0079960

Series editor

Janusz Kacprzyk, Polish Academy of Sciences, Systems Research Institute, Warsaw, Poland

SpringerBriefs present concise summaries of cutting-edge research and practical applications across a wide spectrum of fields. Featuring compact volumes of 50–125 pages, the series covers a range of content from professional to academic.

Typical publications can be:

- A timely report of state-of-the art methods
- An introduction to or a manual for the application of mathematical or computer techniques
- A bridge between new research results, as published in journal articles
- A snapshot of a hot or emerging topic
- An in-depth case study
- A presentation of core concepts that students must understand in order to make independent contributions

SpringerBriefs are characterized by fast, global electronic dissemination, standard publishing contracts, standardized manuscript preparation and formatting guidelines, and expedited production schedules.

On the one hand, **SpringerBriefs in Applied Sciences and Technology** are devoted to the publication of fundamentals and applications within the different classical engineering disciplines as well as in interdisciplinary fields that recently emerged between these areas. On the other hand, as the boundary separating fundamental research and applied technology is more and more dissolving, this series is particularly open to trans-disciplinary topics between fundamental science and engineering.

Indexed by EI-Compendex and Springerlink.

More information about this series at http://www.springer.com/series/8884

Jan-Hendrik Wehner · Dominic Jekel
Rubens Sampaio · Peter Hagedorn

Damping Optimization in Simplified and Realistic Disc Brakes

Springer

Jan-Hendrik Wehner
Weinheim
Germany

Dominic Jekel
Dynamics and Vibrations Group
Technical University of Darmstadt
Darmstadt, Hessen
Germany

Rubens Sampaio
Department of Mechanical Engineering
Pontifical Catholic University of Rio
Rio de Janeiro
Brazil

Peter Hagedorn
Dynamics and Vibration Group
Technical University of Darmstadt
Darmstadt, Hessen
Germany

ISSN 2191-530X ISSN 2191-5318 (electronic)
SpringerBriefs in Applied Sciences and Technology
ISBN 978-3-319-62712-0 ISBN 978-3-319-62713-7 (eBook)
DOI 10.1007/978-3-319-62713-7

Library of Congress Control Number: 2017946042

Printed on acid-free paper

This Springer imprint is published by Springer Nature
The registered company is Springer International Publishing AG
The registered company address is: Gewerbestrasse 11, 6330 Cham, Switzerland

Acknowledgements

The support through DFG HA 1060/55-1, Ingenieurgesellschaft für technische Software mbH (INTES), and Dr.-Ing. h.c. F. Porsche AG is gratefully acknowledged.

Contents

Chapter 1
Introduction

In mechanical engineering, self-excited vibrations are usually unwanted and sometimes dangerous. A famous example is the squealing noise of automotive disc brake systems. The physical origin of this phenomenon is due to friction forces in the contact interface between the brake pads and the rotating brake disc forcing the system to barely visible but audible high frequency vibrations [1]. However, maybe contrary to the public opinion, brake squeal is no indicator of a faulty brake. Even if the squealing noise is a nuisance, the brake system may be in a perfect technical condition with the braking performance ensured [2]. Nevertheless, the automotive industry aims to avoid the occurrence of high-frequency vibrations for comfort reasons. Therefore, finite element (FE) models are used to study brakes under different aspects including the influence of damping on squeal noise [3, 4]. These models may have several hundred thousand or even millions of degrees of freedom (DOF) and are about to reflect the reality in a reasonable manner.

It is well known that the structure of the damping matrix plays an important role in self-excited mechanical systems [5, 6]. Since the physics of damping is the most unknown part in large finite element models, damping is often assumed to be RAYLEIGH damping, i.e. $\mathbf{D} = \alpha\mathbf{M} + \beta\mathbf{K}$. Although this type of modeling the damping matrix is not done on physical grounds but keeps the eigenvectors of the **MDK**-system real, experimental tests identify parameters α and β describing the damping according to different situations. However, since a realistic automotive brake system contains various components and damping origins, RAYLEIGH damping is insufficient to identify the effects of damping for each component individually and stability analyses carried out with this approach are questionable [7].

In this paper, a simple minimal model of disc brake with two DOF is used to gain insight into the different physical origins of damping. Furthermore, FE models of simplified and realistic disc brakes derived by the commercial software package PERMAS are investigated aiming to optimize the damping properties with regard to an equilibrium position which is as stable as possible subject to sensible constraints.

© The Authors 2018
J.-H. Wehner et al., *Damping Optimization in Simplified and Realistic Disc Brakes*, SpringerBriefs in Applied Sciences and Technology,
DOI 10.1007/978-3-319-62713-7_1

In PERMAS, damping can be modeled by different types of linear damping and each component can be treated individually. This, for example, enables assessing the efficiency of anti-squeal shims, which are often applied in automotive brakes [8, 9]. In addition, it is possible to identify those components being more or less worthwhile to be damped in order to avoid brake squeal. Even components where reducing damping has a stabilizing effect may be found, which sometimes can be observed in experimental tests [10]. Consequently, it is not always purposeful to add damping but in some cases it must be reduced to optimize the stability behavior, which may be counterintuitive from an engineer's perspective.

References

1. Kinkaid, N.M., O'Reilly, O.M., Papadopoulos, P.: Automotive disc brake squeal. J. Sound Vib. **267**(1), 105–166 (2003)
2. Breuer, B., Bill, K.H. (eds.): Bremsenhandbuch: Grundlagen, Komponenten, Systeme, Fahrdynamik. ATZ/MTZ-Fachbuch, 4th edn. Springer Vieweg, Wiesbaden (2013)
3. Fritz, G., Sinou, J.-J., Duffal, J.-M., Jézéquel, L.: Effects of damping on brake squeal coalescence patterns—application on a finite element model. Mech. Res. Commun. **34**(2), 181–190 (2007)
4. Fritz, G., Sinou, J.-J., Duffal, J.-M., Jézéquel, L.: Investigation of the relationship between damping and mode-coupling patterns in case of brake squeal. J. Sound Vib. **307**(3–5), 591–609 (2007)
5. Hagedorn, P., Heffel, E., Lancaster, P., Müller, P.C., Kapuria, S.: Some recent results on MDGKN-systems. Z. Angew. Math. Mech. **95**(7), 695–702 (2015)
6. Jekel, D., Hagedorn, P.: Stability of weakly damped MDGKN-systems: the role of velocity proportional terms. Z. Angew. Math. Mech. (1–8) (2017)
7. Hagedorn, P., Eckstein, M., Heffel, E., Wagner, A.: Self-excited vibrations and damping in circulatory systems. J. Appl. Mech. **81**(10), 101009–1–9 (2014)
8. Festjens, H., Gaël, C., Franck, R., Jean-Luc, D., Remy, L.: Effectiveness of multilayer viscoelastic insulators to prevent occurrences of brake squeal: a numerical study. Appl. Acoust. **73**(11), 1121–1128 (2012)
9. Kang, J.: Finite element modelling for the investigation of in-plane modes and damping shims in disc brake squeal. J. Sound Vib. **331**(9), 2190–2202 (2012)
10. Massi, F., Giannini, O.: Effect of damping on the propensity of squeal instability: an experimental investigation. J. Acoust. Soc. Am. **123**(4), 2017–2023 (2008)

Chapter 2
Theoretical Background

2.1 Linearization of Nonlinear Equations of Motion

The equations of motion of a dynamical system in general are nonlinear. Since an analytic solution can only be found for some special cases, it is common practice to linearize these equations around a reference position, typically an equilibrium. In NEWTONIAN mechanics, nonlinear equations of motion can be written in minimal coordinates $\mathbf{q}(t) \in D \subseteq \mathbb{R}^n$ as

$$\mathbf{M}(\mathbf{q}, t)\ddot{\mathbf{q}} + \mathbf{f}(\mathbf{q}, \dot{\mathbf{q}}, t) = \mathbf{0}, \quad \mathbf{q}(0) = \mathbf{q}_0, \quad \dot{\mathbf{q}}(0) = \dot{\mathbf{q}}_0, \tag{2.1}$$

where $\mathbf{M} \in \mathbb{R}^{n \times n}$ is the mass matrix, $(\mathbf{q}(t), t) \in D \times [0, \infty)$ and the function $\mathbf{f} : D \to \mathbb{R}^n \times [0, \infty)$ can be interpreted as the sum of generalized (nonlinear) forces [1]. The number of DOF is equal to the dimension n of the system. Defining $\Delta \mathbf{q}(t) = \mathbf{q}(t) - \mathbf{q}_e(t)$ as a (small) deviation from an equilibrium point $\mathbf{q}_e$, that may be put to zero without loss of generality, a TAYLOR expansion yields the linearized equations of motion

$$\mathbf{M}(t)\ddot{\mathbf{q}} + \mathbf{B}(t)\dot{\mathbf{q}} + \mathbf{C}(t)\mathbf{q} = \mathbf{g}(t). \tag{2.2}$$

Looking for a physical meaning of $\mathbf{B}(t)$ and $\mathbf{C}(t)$, it is advantageous to decompose the matrices into their symmetric and skew-symmetric parts

$$\mathbf{D}(t) = \frac{1}{2} \left(\mathbf{B}(t) + \mathbf{B}(t)^{\mathrm{T}} \right), \qquad \mathbf{G}(t) = \frac{1}{2} \left(\mathbf{B}(t) - \mathbf{B}(t)^{\mathrm{T}} \right),$$
$$\mathbf{K}(t) = \frac{1}{2} \left(\mathbf{C}(t) + \mathbf{C}(t)^{\mathrm{T}} \right), \qquad \mathbf{N}(t) = \frac{1}{2} \left(\mathbf{C}(t) - \mathbf{C}(t)^{\mathrm{T}} \right), \tag{2.3}$$

where $\mathbf{D}(t)$ is the damping matrix, $\mathbf{G}(t)$ is the gyroscopic matrix, $\mathbf{K}(t)$ is the stiffness matrix and $\mathbf{N}(t)$ is the matrix containing circulatory terms [2]. Often homogenous systems are considered, i.e. setting $\mathbf{g}(t) \equiv \mathbf{0}$ yields the **MDGKN**-system

© The Authors 2018
J.-H. Wehner et al., *Damping Optimization in Simplified and Realistic Disc Brakes*, SpringerBriefs in Applied Sciences and Technology, DOI 10.1007/978-3-319-62713-7_2

$$\mathbf{M}(t)\ddot{\mathbf{q}}(t) + [\mathbf{D}(t) + \mathbf{G}(t)]\dot{\mathbf{q}}(t) + [\mathbf{K}(t) + \mathbf{N}(t)]\mathbf{q}(t) = \mathbf{0} \qquad (2.4)$$

describing free linear vibrations around an equilibrium position.

2.2 Time-Invariant MDGKN-Systems

For the time being, the matrices are assumed to be time-invariant, i.e.

$$\mathbf{M}\ddot{\mathbf{q}}(t) + (\mathbf{D} + \mathbf{G})\dot{\mathbf{q}}(t) + (\mathbf{K} + \mathbf{N})\mathbf{q}(t) = \mathbf{0}. \qquad (2.5)$$

Looking for a solution of the form $\mathbf{q}(t) = \hat{\mathbf{q}}e^{\lambda t}$ yields the eigenvalue problem, where $\lambda \in \mathbb{C}$ is an eigenvalue and $\hat{\mathbf{q}} \in \mathbb{C}^n$ is the corresponding (right) eigenvector. The imaginary part of λ represents the circular frequency ω, or, if divided by 2π, the frequency of the corresponding mode, i.e. $\lambda = \mathrm{Re}(\lambda) \pm \mathrm{i}\omega$. Nontrivial solutions only exist if

$$\det[\lambda^2 \mathbf{M} + \lambda(\mathbf{D} + \mathbf{G}) + \mathbf{K} + \mathbf{N}] = 0. \qquad (2.6)$$

After determining the eigenvalues λ_i ($i = 1, \ldots, 2n$), the eigenvectors $\hat{\mathbf{q}}_i$ can be calculated as a solution of $[\lambda_i^2 \mathbf{M} + \lambda_i(\mathbf{D} + \mathbf{G}) + \mathbf{K} + \mathbf{N}]\hat{\mathbf{q}}_i = 0$. Since (2.5) is linear, the general solution

$$\mathbf{q}(t) = \sum_{i=1}^{2n} C_i \hat{\mathbf{q}}_i e^{\lambda_i t} \qquad (2.7)$$

can be found by superposition certainly as long as there are no multiple eigenvalues. The constants $C_i \in \mathbb{C}$ are determined by the initial conditions $\mathbf{q}_0$ and $\dot{\mathbf{q}}_0$ given in (2.1).

With respect to stability analysis, the real part of the eigenvalues λ_i decides whether the absolute value of (2.7) grows or decays in time. Therefore, in the linear case, the following conditions for LYAPUNOV stability can be defined [3]:

(1) If each λ_i ($i = 1, \ldots, 2n$) has negative real part, i.e. $\mathrm{Re}(\lambda_i) < 0 \ \forall \ i \in \{1, \ldots, 2n\}$, the trivial solution is *asymptotically stable*.
(2) If at least one eigenvalue has positive real part, i.e. $\exists \ i \in \{1, \ldots, 2n\} : \mathrm{Re}(\lambda_i) > 0$, the trivial solution is *unstable*.

If neither condition 1 nor 2 is satisfied, a critical case occurs and the stability of the equilibrium depends on nonlinear terms.

2.3 First-Order Systems

The coordinate transformation $\mathbf{y}(t) = (\mathbf{q}(t), \dot{\mathbf{q}}(t))^{\mathsf{T}}$ with $\mathbf{y} \in \mathbb{R}^{2n}$ yields a first-order formulation of (2.5), i.e.

$$\dot{\mathbf{y}}(t) = \underbrace{\begin{bmatrix} \mathbf{0} & \mathbf{I} \\ -\mathbf{M}^{-1}(\mathbf{K}+\mathbf{N}) & -\mathbf{M}^{-1}(\mathbf{D}+\mathbf{G}) \end{bmatrix}}_{:=\mathbf{A}} \mathbf{y}(t), \tag{2.8}$$

where $\mathbf{A} \in \mathbb{R}^{2n \times 2n}$ is the square coefficient matrix and $\mathbf{I}$ is the $n \times n$ identity matrix. The eigenvalues defined by (2.6) are identical to those of $\mathbf{A}$. Any set of $2n$ linearly independent solutions of (2.8) is called a *fundamental system*. The term *fundamental system* is not restricted to linear systems. But, in most cases, it can be found analytically for linear systems only. The *fundamental matrix* $\boldsymbol{\Phi}(t)$ is defined as

$$\boldsymbol{\Phi}(t) := (\mathbf{y}_1(t) \mid \mathbf{y}_2(t) \mid ... \mid \mathbf{y}_{2n}(t)) \in \mathbb{R}^{2n \times 2n}. \tag{2.9}$$

It can also be determined by using the matrix exponential according to [4]. Let $\mathbf{X}$ be an arbitrary square matrix of arbitrary dimension, the *matrix exponential* is defined as

$$e^{\mathbf{X}} = \sum_{k=0}^{\infty} \frac{1}{k!} \mathbf{X}^k. \tag{2.10}$$

The *fundamental matrix* of the first-order system $\dot{\mathbf{y}}(t) = \mathbf{A}\mathbf{y}(t)$ is then given by

$$\boldsymbol{\Phi}(t) = \boldsymbol{\Phi}(0) e^{\mathbf{A}t}. \tag{2.11}$$

Setting the initial values $\boldsymbol{\Phi}(0) = \mathbf{I}$, Eq. (2.11) simplifies to

$$\boldsymbol{\Phi}(t) = e^{\mathbf{A}t}. \tag{2.12}$$

2.4 Time-Periodic Systems and FLOQUET Theory

In the previous sections, two ways of analyzing and finding solutions for time-invariant **MDGKN**-systems are discussed. In many technical applications, e.g. automotive brakes, the assumption of time-invariance yields satisfying results. However, for time-variant linear systems, in general no solution can be found analytically. Assuming time-periodic coefficients, conclusions about the stability behavior can still be drawn. Consider the linear first-order system

$$\dot{\mathbf{y}}(t) = \mathbf{A}(t)\mathbf{y}(t), \ \mathbf{A}(t) = \mathbf{A}(t + \frac{2\pi}{\Omega}), \tag{2.13}$$

where, similar to the time-invariant case, $\mathbf{A}(t) \in \mathbb{R}^{2n \times 2n}$. In (2.8), for example, the matrix $\mathbf{K}(t)$ may be periodic. Using FLOQUET theory the *fundamental matrix* of system (2.13) can be written as

$$\mathbf{\Phi}(t) = \mathbf{P}_\Omega(t)e^{\mathbf{B}t},\tag{2.14}$$

where $\mathbf{P}_\Omega(t) = \mathbf{P}_\Omega(t + \frac{2\pi}{\Omega})$ is a $\frac{2\pi}{\Omega}$-periodic matrix [5]. The index Ω denotes an angular velocity forcing the system to time-periodic parameters. Since system (2.13) repeats its properties periodically, it is sufficient to examine the stability behavior over one period Ω only. If the solution grows after one period, the system is unstable and vice versa. Hence, the main interest is not $\mathbf{\Phi}(t)$ but $\mathbf{\Phi}(T)$, the *fundamental matrix* at the time T after one period. The *monodromy matrix*

$$\mathcal{M} := e^{T\mathbf{B}}\tag{2.15}$$

contains information about the state after one period [5, 9]. If the initial conditions are $\mathbf{\Phi}(0) = \mathbf{I}$, the *monodromy matrix* is equal to the *fundamental matrix* evaluated after one period. The eigenvalues of $\mathcal{M}$ are called *Floquet multipliers* and are denoted by $\mu_i \in \mathbb{C}, i = 1, \ldots, 2n$. Similar to the time-invariant case, the following stability conditions can be defined [5]:

(1) If the magnitude of each *Floquet multiplier* is smaller than one, i.e. $|\mu_i| < 1 \, \forall \, i \in [1, \ldots, 2n]$, system (2.13) is *asymptotically stable*.
(2) If the magnitude of at least one *Floquet multiplier* is larger than one, i.e. $\exists \, i \in [1, \ldots, 2n] : |\mu_i| > 1$, system (2.13) is *unstable*.

The solution of time-periodic systems can be written as

$$\mathbf{\Phi}(t) = e^{\int_0^t \mathbf{A}(\tau)d\tau},\tag{2.16}$$

where the initial conditions are set to the identity matrix. Equation (2.16) is a generalized version of (2.12), which is valid for $\mathbf{A}(t)$ being periodic or not. Since the integral of the time-dependent matrix cannot be solved analytically in general, the solution can only be computed numerically at specified time intervals. FLOQUET theory predicts $T = \frac{2\pi}{\Omega}$ as the time interval to be sufficient to study the stability behavior of (2.13). This yields another formulation of the *monodromy matrix*

$$\mathcal{M} = e^{\int_0^T \mathbf{A}(\tau)d\tau}.\tag{2.17}$$

For solving the integral in (2.17) numerically, T can be divided into m subintervals, i.e. $\Delta t = \frac{T}{m}$. Discrete evaluation leads to

$$\mathcal{M} = \lim_{m \to \infty} \prod_{j=0}^{m} e^{\mathbf{A}(j\Delta t)\Delta t}, \quad \Delta t = \frac{T}{m} = \frac{2\pi}{\Omega m}.\tag{2.18}$$

Choosing m sufficiently large, Eq. (2.18) finally gives an approximation for the *monodromy matrix* [6]. For (2.18) to be evaluated, an explicit expression of $\mathbf{A}(t)$ has to be be given.

2.5 Optimization of Damping

In this study, two kinds of linear systems are considered. The decision whether the system is stable or unstable either depends on the real part of λ in the time-invariant case or on the magnitude of μ in the time-periodic case. Instability occurs if at least one eigenvalue has positive real part or if at least one eigenvalue of the *monodromy matrix* has absolute value greater than one, respectively. Thus, the optimization of the stability behavior leads to an optimization of eigenvalues $\lambda_i = \lambda_i(\mathbf{p})$ or *Floquet multipliers* $\mu_i = \mu_i(\mathbf{p})$, respectively, where $\mathbf{p} \in \mathbb{R}^s$ is a vector containing s parameters.

2.5.1 Time-Invariant Systems

As discussed in [7], the optimization of time-invariant systems can be formulated as

$$
\begin{aligned}
&\min_{\hat{\mathbf{p}}} \max_{i} \operatorname{Re}(\lambda_i) \\
&\text{s.t. } \boldsymbol{\Gamma}\hat{\mathbf{p}} \leq \mathbf{c}, \\
&\boldsymbol{\Gamma} \in \mathbb{R}^{r \times \hat{s}}, \ \hat{\mathbf{p}} \in \mathbb{R}^{\hat{s}} \subseteq \mathbf{p}, \ \mathbf{c} \in \mathbb{R}^{r}, \ \hat{s} \leq s.
\end{aligned}
\tag{2.19}
$$

Here, $\boldsymbol{\Gamma}\hat{\mathbf{p}} \leq \mathbf{c}$ represents a set of r linear equations constraining $\hat{\mathbf{p}}$ in an admissible range (polyhedron). Since not all parameters are optimized in general, $\hat{\mathbf{p}}$ contains $\hat{s}$ parameters only.

Another possibility to improve the stability is to maximize the minimum damping ratio

$$
D_i = -\frac{\operatorname{Re}(\lambda_i)}{|\lambda_i|}
\tag{2.20}
$$

according to [8]. In this case, the first line in (2.19) changes to $\max_{\hat{\mathbf{p}}} \min_{i} D_i(\lambda_i)$.

2.5.2 Time-Periodic Systems

For time-periodic systems, the formulation of the optimization problem is similar, i.e.

$$\min_{\hat{\mathbf{p}}} \max_{i} |\mu_i|$$

$$\text{s.t. } \mathbf{\Gamma}\hat{\mathbf{p}} \leq \mathbf{c}, \tag{2.21}$$

$$\mathbf{\Gamma} \in \mathbb{R}^{r \times \hat{s}}, \ \hat{\mathbf{p}} \in \mathbb{R}^{\hat{s}} \subseteq \mathbf{p}, \ \mathbf{c} \in \mathbb{R}^r, \ \hat{s} \leq s.$$

In this case, the maximum magnitude of the *Floquet multipliers* has to be minimized. To define a damping ratio similar to (2.20), the relationship

$$\mu_i = e^{\lambda_i T} \tag{2.22}$$

connecting the eigenvalues of the time-invariant system with the *Floquet multipliers* is used [9]. With (2.22) it is possible to determine a formula for the real part of λ_i, i.e.

$$\text{Re}(\lambda_i) = \frac{1}{T}\ln|\mu_i|. \tag{2.23}$$

However, the imaginary part of a *Floquet multiplier* does not really represent circular frequencies, which may be strongly influenced by the function $P_\Omega(t)$. Still, to define a damping ratio for parametrically excited systems, this problem can be solved by determining numerically the dominant frequency of (2.16) with a discrete FOURIER analysis. The calculation is carried out with the fast FOURIER transformation (FFT) implemented in MATLAB. A detailed explanation of this algorithm can be found in [10]. The equivalent damping ratio becomes

$$\tilde{D}_i = -\frac{\ln|\mu_i|}{T\sqrt{(\frac{1}{T}\ln|\mu_i|)^2 + (2\pi f_i)^2}}, \tag{2.24}$$

where f_i denotes the frequency of $\mathbf{q}_i$ resulting from the FFT. Thus, the first line in (2.21) changes to $\max_{\hat{\mathbf{p}}} \min_{i} \tilde{D}_i(\mu_i)$.

2.6 Linear Damping Models

Regarding optimization problems (2.19) and (2.21), the main focus of this study is to modify the properties of the damping matrix $\mathbf{D}$ in order to stabilize or make more stable the equilibrium state subject to sensible constraints. Since most physical processes and in particular damping are inherently nonlinear, the damping models are linearized. Consider the forced one-degree-of-freedom system

$$m\ddot{q} + d_{eq}\dot{q} + kq = \hat{f}e^{i\Omega t}, \tag{2.25}$$

where d_{eq} is the equivalent viscous damping constant of the respective model. In contrast to the homogenous case, where the right-hand side of (2.25) is equal to zero, a particular solution of (2.25) of the form

$$q = \hat{q} e^{i\Omega t + \alpha t} \tag{2.26}$$

is postulated for the definition of structural damping. The dissipated energy E_D caused by a damping force F_D over one period T is given by

$$E_D = \int_0^T F_D \dot{q} \, dq. \tag{2.27}$$

In the following, COULOMB damping (damping constant $d_{eq,c}$), viscous damping (damping constant $d_{eq,v}$) and structural damping (damping constant $d_{eq,m}$) are discussed.

2.6.1 COULOMB *Damping*

The friction force $F_{D,c}$ between two bodies sliding with respect to each other is given by COULOMB's law, i.e.

$$F_{D,c} = \frac{\mu|N|}{|\dot{q}|}\dot{q}, \quad \dot{q} \neq 0, \tag{2.28}$$

where μ is the friction coefficient, N is the normal force and $\dot{q}$ is the relative velocity between the two bodies [11]. According to COULOMB, the friction coefficient is not a function of q, $\dot{q}$ and N and is assumed to be constant. A coefficient comparison between (2.25) and (2.28) yields the equivalent viscous damping coefficient of a system with COULOMB damping, i.e. $d_{eq,c} = \frac{\mu|N|}{|\dot{q}|}$. As can be seen in Fig. 2.1, the homogenous COULOMB damped solution of (2.25) with $\hat{f} = 0$ decays linearly in time.

Fig. 2.1 Amplitude of solution in COULOMB damped system

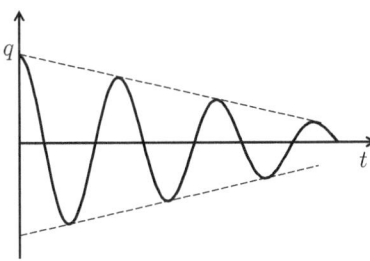

2.6.2 Viscous Damping

The resulting force of viscous damping $F_{D,c}$ is proportional to the relative velocity $\dot{q}$, i.e.

$$F_{D,v} = d\dot{q}. \tag{2.29}$$

A coefficient comparison with (2.25) yields $d_{eq,v} = d$, where d is the viscous damping coefficient. If the damping force does not only depend on the current velocity but also on the history of the velocity, Eq. (2.29) becomes the convolution integral

$$F_{D,v}(t) = \int_{-\infty}^{\infty} b(t - \tau)\dot{q}(\tau)d\tau, \tag{2.30}$$

where the function $b(t - \tau)$ relates $q(\tau)$ and $F_v(t)$. Since viscous damping cannot depend on the velocity in the future, the *causality condition*

$$b(t - \tau) \overset{!}{=} 0, \quad \text{if } \tau > t, \tag{2.31}$$

has to be satisfied. Inserting (2.26) in (2.29) and using (2.27) yields the dissipated energy over one period

$$E_{D,v} \propto \Omega d |\hat{q}|^2 \tag{2.32}$$

which is proportional to the square of the amplitude and proportional to the circular frequency. As can be seen in Fig. 2.2, the solution of a viscously damped homogenous system decays exponentially in time.

2.6.3 Structural Damping

Experiments on structural elements and also on complex structures indicate that the energy dissipated internally in cyclic straining is often proportional to the square of the amplitude but independent of the frequency Ω [12]. A suitable representation

Fig. 2.2 Amplitude of solution in viscously damped system

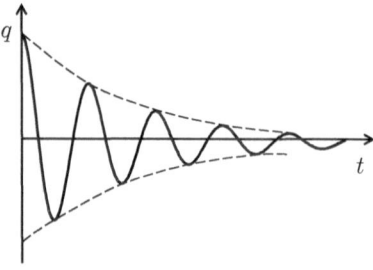

Fig. 2.3 KELVIN- VOIGT model

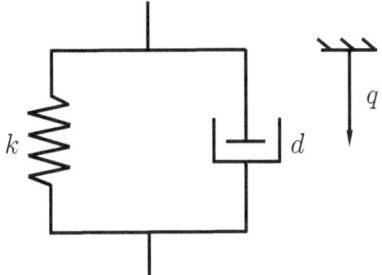

of material behavior is the KELVIN- VOIGT model shown in Fig. 2.3. The elastic spring (stiffness k) and the viscous damper (damping coefficient d) are connected in parallel. The resulting force $F_{D,m} = kq + d\dot{q}$ is the sum of spring and damper force [13]. Assuming harmonic oscillations according to (2.26) yields the complex stiffness notation

$$F_{D,m} = k(1 + ig)q, \qquad (2.33)$$

where $g := \frac{d\Omega}{k}$ is the structural damping factor (loss factor) which has to be determined experimentally [14]. Detailed explanation of the experiments can be found in [15]. The dissipated energy over one period is then

$$E_{D,m} \propto gk|\hat{q}|^2. \qquad (2.34)$$

Using (2.32) and (2.34) the equivalent viscous damping factor $d_{\mathrm{eq},m} = \frac{gk}{\Omega_{\mathrm{ref}}}$ can be defined. As can be seen in Fig. 2.4, Ω_{ref} is the circular frequency at $E_{D,v} = E_{D,m}$. If $\Omega > \Omega_{\mathrm{ref}}$, structural damping is overestimated and vice versa.

In general, the definition of structural damping is valid only for single-degree-of-freedom systems harmonically excited with one frequency [14]. In homogeneous systems, i.e. $\hat{f}e^{i\Omega t} \equiv 0$, Eq. (2.25) requires a solution of the form $q = \hat{q}e^{-\delta t + i(\omega_d + \gamma t)}$, where $\delta = \frac{d}{m}$ is the constant of exponential decay, ω_d is the frequency of the damped oscillation and γ is the phase shift. In this case, the resulting damping force is

Fig. 2.4 Reference circular frequency

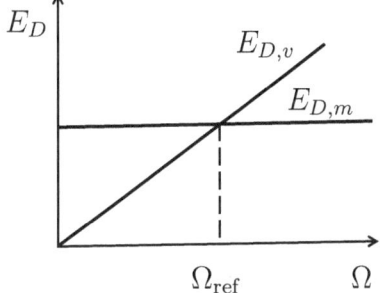

$$F_{D,m} = k \left(\sqrt{1 - g^2} + ig \right) q. \tag{2.35}$$

In many technically relevant applications, the structural damping factor is taken to be $\mathcal{O}(h^{-2})$ or smaller [16]. With this assumption ($g \ll 1$), Eqs. (2.33) and (2.35) are approximately equal and the approach can be extended to systems with $n > 1$. Then, the complex stiffness and the equivalent viscous damping matrix read

$$\mathbf{K}^* = \mathbf{K} + i\mathbf{H}, \quad \mathbf{D}_S = \frac{1}{\Omega_{\text{ref}}} \mathbf{H}, \tag{2.36}$$

where the structural damping matrix is $\mathbf{H} := g\mathbf{K}$. A limit of this damping model is that a loss factor g independent of the circular frequency violates the *causality condition* (2.31) in the time-domain [17, 18]. However, since many problems in vibration theory are analyzed in the frequency domain and only within certain limited frequency bands, the approach remains feasible.

2.7 Modal Reduction

Mechanical systems discretized using finite elements have a large number of DOF. In order to save computing time modal reduction techniques are frequently used to reduce the number of DOF while maintaining the main dynamical characteristics of the system. In a linear combination of n eigenvectors $\hat{\mathbf{q}}_k \in \mathbb{R}^n$ the vector of minimal coordinates

$$\mathbf{q} = \sum_{k=1}^{n} \hat{\mathbf{q}}_k p_k = \mathbf{X}\mathbf{p} \tag{2.37}$$

can be expressed exactly, where p_k are the modal coordinates and $\mathbf{X} \in \mathbb{R}^{n \times n}$ is the modal matrix. In modal reduction, the vector of minimal coordinates is approximated by only $m \ll n$ eigenvectors, i.e.

$$\mathbf{q} \approx \sum_{k=1}^{m} \hat{\mathbf{q}}_k p_k = \mathbf{X}_m \mathbf{p}, \tag{2.38}$$

where $\mathbf{X}_m \in \mathbb{R}^{n \times m}$ is a rectangular matrix transforming an n-degree-of-freedom system into a subspace containing only m DOF. In order to determine the most suitable eigenvectors, different modal reduction techniques have been developed [19, 20]. The equations of motion in modal space read

$$\tilde{\mathbf{M}}\ddot{\mathbf{p}} + (\tilde{\mathbf{D}} + \tilde{\mathbf{G}})\dot{\mathbf{p}} + (\tilde{\mathbf{K}} + \tilde{\mathbf{N}})\mathbf{p} = \mathbf{0}, \tag{2.39}$$

where $\tilde{\mathbf{M}} = \mathbf{X}_m^T \mathbf{M} \mathbf{X}_m$, $\tilde{\mathbf{D}} = \mathbf{X}_m^T \mathbf{D} \mathbf{X}_m$, $\tilde{\mathbf{G}} = \mathbf{X}_m^T \mathbf{G} \mathbf{X}_m$, $\tilde{\mathbf{K}} = \mathbf{X}_m^T \mathbf{K} \mathbf{X}_m$, and $\tilde{\mathbf{N}} = \mathbf{X}_m^T \mathbf{N} \mathbf{X}_m$ are the respective truncated modal matrices.

2.8 Brake Squeal

A first physical explanation of brake squeal is based on the assumption of a decreasing friction coefficient, i.e. by a negative slope of the friction-velocity characteristic [21]. However, experimental tests showed the phenomenon not to be the result of oscillation components in circumferential direction [22]. Instead, it is now commonly accepted that brake squeal is initiated by an instability due to friction forces leading to self-excited vibrations even for constant coefficients of friction. These friction forces lead to an asymmetry in the matrices describing the coordinate proportional forces and to mode coupling effects which may lead to flutter-type instabilities [23, 24]. In case of nonlinear models, the disc oscillates ultimately reaching a limit cycle. Detailed explanations of the nonlinear case can be found in [11, 25–27]. Since brake squeal is due to unstable solutions, in the linear case, positive real parts of the eigenvalues λ may lead to squeal. Therefore, complex eigenvalue analysis (CEA) is the approach mostly favoured by the automotive industry [28]. However, the magnitude of the eigenvalue real part (or damping ratio) does not predict the relative amplitude, i.e. the sound pressure level. It only reveals how fast the vibration is growing and thus represents only a relative measure of squeal propensity. For example, a linear analysis can predict an unstable system, but the resulting limit cycle may be very small such that the noise generated would be inaudible.

Different techniques have been developed to avoid the occurrence of the noise phenomenon. WAGNER showed that splitting the eigenfrequencies, which is implemented using an optimization of the geometric properties of the disc, minimizes the tendency of the disc to squeal [29]. Similar results are provided by SPELSBERG-KORSPETER [30]. Another technical approach is an active control of the brake pads using piezoceramic actuators [31]. In [32], the main goal is to stabilize the disc by optimizing parameter values of the damping matrix using a minimal model of disc brake developed by VON WAGNER et al. in [33]. In another work, it is shown that damping properties, especially the damping ratio of different modes, have to be considered in detail in order to avoid mode coupling and friction induced vibrations [34].

References

1. Seyranian, A.P., Mailybaev, A.A.: Multiparameter Stability Theory with Mechanical Applications. Series on Stability, Vibration, and Control of Systems, vol. 13. Series A. World Scientific, Singapore (2003)
2. Müller, P.C.: Stabilität und Matrizen: Matrizenverfahren in der Stabilitätstheorie linearer dynamischer Systeme. Ingenieurwissenschaftliche Bibliothek. Springer, Berlin, Heidelberg, New York (1977)
3. Kuypers, F.: Klassische Mechanik, 10th edn. Wiley-VCH, Weinheim (2016)
4. Horn, R.A., Johnson, C.R.: Topics in Matrix Analysis. University Press, Cambridge (1995)
5. Chicone, C.: Ordinary Differential Equations with Applications. Texts in Applied Mathematics, vol. 34, 2nd edn. Springer, New York (2006)
6. Spelsberg-Korspeter, G.: Eigenvalue optimization against brake squeal: symmetry, mathematical background and experiments. J. Sound Vib. **331**(19), 4259–4268 (2012)

7. Jekel, D., Clerkin, E., Hagedorn, P.: Robust damping in self-excited mechanical systems. Proc. Appl. Math. Mech. **16**(1), 695–696 (2016)
8. Müller, P.C., Schiehlen, W.O.: Lineare Schwingungen: Theoretische Behandlung von mehrfachen Schwingern. Akademische Verlagsgesellschaft, Wiesbaden (1976)
9. Gasch, R., Knothe, K., Liebich, R.: Strukturdynamik: Diskrete Systeme und Kontinua, 2nd edn. Springer, Berlin, Heidelberg (2012)
10. Arens, T., Hettlich, F., Karpfinger, C., Kockelkorn, U., Lichtenegger, K., Stachel, H.: Mathematik. Springer Spektrum, Berlin and Heidelberg (2015)
11. Hochlenert, D.: Selbsterregte Schwingungen in Scheibenbremsen: Mathematische Modellbildung und aktive Unterdrückung von Bremsenquietschen. Berichte aus dem Maschinenbau. Shaker, Aachen (2006)
12. Kimball, A.L., Lovell, D.E.: Internal friction in solids. Phys. Rev. **30**(6), 948–959 (1927)
13. Mezger, T.G.: Das Rheologie Handbuch: Für Anwender von Rotations- und Oszillations-Rheometern. Farbe-und-Lack-Bibliothek, 5th edn. Vincentz, Hannover (2016)
14. Neumark, S.: Concept of Complex Stiffness Applied to Problems of Oscillations with Viscous and Hysteretic Damping. Technical Report 3269, Aeronautical Research Council, London (1962)
15. Bert, C.W.: Material damping. J. Sound Vib. **29**(2), 129–153 (1973)
16. Stevenson, J.D.: Structural damping values as a function of dynamic response stress and deformation levels. Nucl. Eng. Des. **60**(2), 211–237 (1980)
17. Crandall, S.H.: The role of damping in vibration theory. J. Sound Vib. **11**(1), 3–18 (1970)
18. Woodhouse, J.: Linear damping models for structural vibration. J. Sound Vib. **215**(3), 547–569 (1998)
19. Antoulas, A.C., Sorensen, D.C., Gugercin, S.: A survey of model reduction methods for large-scale systems. Contemp. Math. **280**, 193–220 (2001)
20. Qu, Z.-Q.: Model Order Reduction Techniques: With Applications in Finite Element Analysis. Springer, London (2004)
21. Ouyang, H., Mottershead, J.E., Cartmell, M.P., Friswell, M.I.: Friction-induced parametric resonances in discs: effect of a negative friction-velocity relationship. J. Sound Vib. **209**(2), 251–264 (1998)
22. Hetzler, H.: Zur Stabilität von Systemen bewegter Kontinua mit Reibkontakten am Beispiel des Bremsenquietschens. Schriftenreihe des Instituts für Technische Mechanik, vol. 8. Univ.-Verl. Karlsruhe, Karlsruhe (2008)
23. Hoffmann, N., Gaul, L.: Effects of damping on mode-coupling instability in friction induced oscillations. Z. Angew. Math. Mech. **83**(8), 524–534 (2003)
24. Sinou, J.-J., Jézéquel, L.: Mode coupling instability in friction-induced vibrations and its dependency on system parameters including damping. Euro. J. Mech. A Solids **26**(1), 106–122 (2007)
25. Massi, F., Baillet, L., Giannini, O., Sestieri, A.: Brake squeal: linear and nonlinear numerical approaches. Mech. Syst. Signal Process. **21**(6), 2374–2393 (2007)
26. Sinou, J.-J.: Transient non-linear dynamic analysis of automotive disc brake squeal—on the need to consider both stability and non-linear analysis. Mech. Res. Commun. **37**(1), 96–105 (2010)
27. Spelsberg-Korspeter, G., Hochlenert, D., Hagedorn, P.: Non-linear investigation of an asymmetric disk brake model. Proc. IMechE Part C: J. Mech. Eng. Sci. **225**(10), 2325–2332 (2011)
28. Ouyang, H., Nack, W.V., Yuan, Y., Chen, F.: Numerical analysis of automotive disc brake squeal: a review. Int. J. Veh. Noise Vib. **1**(3/4), 207–231 (2005)
29. Wagner, A.: Avoidance of brake squeal by a separation of the brake disc's eigenfrequencies: a structural optimization problem. Forschungsbericht, vol. 31. Studienbereich Mechanik, Darmstadt (2013)
30. Spelsberg-Korspeter, G.: Breaking of symmetries for stabilization of rotating continua in frictional contact. J. Sound Vib. **322**(4–5), 798–807 (2009)
31. Schlagner, S., von Wagner, U.: Evaluation of automotive disk brake noise behavior using piezoceramic actuators and sensors. Proc. Appl. Math. Mech. **7**(1), 4050031–4050032 (2007)

32. Jekel, D., Hagedorn, P.: Stability of weakly damped MDGKN-systems: the role of velocity proportional terms. Z. Angew. Math. Mech. (1–8) (2017)
33. von Wagner, U., Hochlenert, D., Hagedorn, P.: Minimal models for disk brake squeal. J. Sound Vib. **302**(3), 527–539 (2007)
34. Sinou, J.-J., Jézéquel, L.: On the stabilizing and destabilizing effects of damping in a non-conservative pin-disc system. Acta Mech. **199**(1–4), 43–52 (2008)

Chapter 3
Optimization of a Minimal Model of Disc Brake

In this study, the latter idea of optimizing the damping matrix is explored further. In a first step, the minimal model of disc brake developed in [1] is analyzed. This model has two DOF so that the results can only give an insight into the stability behavior of an idealized brake.

3.1 Equations of Motion

Consider a rigid disc (moment of inertia Θ, radius r, and thickness h) rotating with constant angular speed Ω around its center of mass with the angles q_1 and q_2 being minimal coordinates, cf. Fig. 3.1. Two pins representing the brake pads can move freely but under the actions of springs (stiffness k) and dampers (damping coefficient d) in the n_3-direction of a NEWTONian reference frame and in frictional contact with the disc. Based on experimental tests conducted in [2] the friction is assumed to be of COULOMB type with a constant and isotropic coefficient μ in a first approximation. Detailed information and experimental results about a velocity dependent friction coefficient can be found in [3]. It is further assumed that there is only slip between the disc and the pins which is assured as long as the rotational speed of the disc is sufficiently large. The prestress in the pins is N_0, the supports of the rotating disc have the stiffness and damping properties k_t and d_t, respectively. The parameters

$$h = 0.02\,\text{m},\ r = 0.13\,\text{m},\ \Omega = 5\pi\,\text{s}^{-1},\ \mu = 0.6,\ N_0 = 3000\,\text{N},\ \Theta = 0.16\,\text{kgm}^2,$$

$$k = 6.00 \times 10^6\,\text{N/m},\ k_t = 1.88 \times 10^7\,\text{Nm},\ d = 5.0\,\text{Ns/m},\ d_t = 0.1\,\text{Nms}$$

$$(3.1)$$

are chosen to compare with [1].

© The Authors 2018

J.-H. Wehner et al., *Damping Optimization in Simplified and Realistic Disc Brakes*, SpringerBriefs in Applied Sciences and Technology, DOI 10.1007/978-3-319-62713-7_3

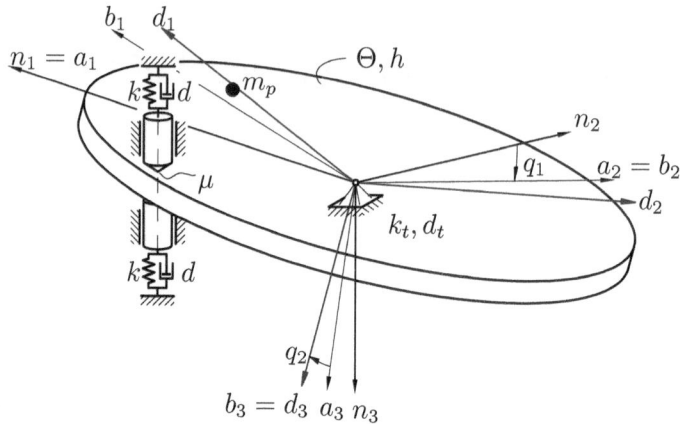

Fig. 3.1   Minimal model of wobbling disc brake

In addition, to bring time-variance into the model, an asymmetric bearing and a mass point m_p with the distance r_p on the body fixed d_1-axis of the disc are assumed. In an actual brake, time-periodicity is due to ventilation channels. The equations of motion then have the perturbation parameters $\kappa = \frac{k_{t2}-k_{t1}}{k_{t1}} > 0$, $\delta = \frac{d_{t2}-d_{t1}}{d_{t1}} > 0$, and $\Theta_p = m_p r_p^2$ leading to time-periodic matrices which may force the system to exhibit parametrically excited vibrations. If the bearing is symmetric, k_t and d_t are independent of direction, i.e. $k_{t1} = k_{t2} = k_t$, $d_{t1} = d_{t2} = d_t$, and hence $\kappa = \delta = 0$. As outlined in [4], the matrices describing the equations of motion (2.4) become

$$\mathbf{M} = \begin{bmatrix} \Theta & 0 \\ 0 & \Theta \end{bmatrix} + \Theta_p \begin{bmatrix} \sin^2(\Omega t) & -\sin(\Omega t)\cos(\Omega t) \\ -\sin(\Omega t)\cos(\Omega t) & \cos^2(\Omega t) \end{bmatrix},$$

$$\mathbf{D} = \begin{bmatrix} d_t(1 + \tfrac{1}{2}\delta) + 2dr^2 + \tfrac{1}{2}\mu N_0 \frac{h^2}{\Omega r} & -\tfrac{1}{2}\mu dhr \\ -\tfrac{1}{2}\mu dhr & d_t(1 + \tfrac{1}{2}\delta) \end{bmatrix} - \Theta_p \Omega \begin{bmatrix} -\sin(2\Omega t) & \cos(2\Omega t) \\ \cos(2\Omega t) & \sin(2\Omega t) \end{bmatrix}$$
$$- \frac{1}{2}\delta d_t \begin{bmatrix} \cos(2\Omega t) & \sin(2\Omega t) \\ \sin(2\Omega t) & -\cos(2\Omega t) \end{bmatrix},$$

$$\mathbf{G} = \begin{bmatrix} 0 & (2\Theta + \Theta_p)\Omega + \tfrac{1}{2}\mu dhr \\ -[(2\Theta + \Theta_p)\Omega + \tfrac{1}{2}\mu dhr] & 0 \end{bmatrix}, \tag{3.2}$$

$$\mathbf{K} = \begin{bmatrix} k_t(1 + \tfrac{1}{2}\kappa) + 2kr^2 + N_0 h & -\tfrac{1}{4}\mu r[2kh + N_0(4 - \frac{h^2}{r^2})] \\ -\tfrac{1}{4}\mu r[2kh + N_0(4 - \frac{h^2}{r^2})] & k_t(1 + \tfrac{1}{2}\kappa) + (1 + \mu^2)N_0 h \end{bmatrix} - \frac{1}{2}\kappa k_t \begin{bmatrix} \cos(2\Omega t) & \sin(2\Omega t) \\ \sin(2\Omega t) & -\cos(2\Omega t) \end{bmatrix},$$

$$\mathbf{N} = \begin{bmatrix} 0 & \tfrac{1}{4}\mu r[2kh + N_0(4 + \frac{h^2}{r^2})] \\ -\tfrac{1}{4}\mu r[2kh + N_0(4 + \frac{h^2}{r^2})] & 0 \end{bmatrix}.$$

Setting $m_p = 0$ the matrices simplify as shown in [4]. In order to get an insight into the different physical origins of damping and to formulate an optimization problem it is advantageous to decompose the damping matrix into $\mathbf{D} = \mathbf{D}_{\text{friction}} + \mathbf{D}_{\text{pad}} + \mathbf{D}_{\text{disc}}(t) + \mathbf{D}_{m_p}(t)$, see [5], where

$$\mathbf{D}_{\text{friction}} = \begin{bmatrix} \frac{1}{2}\mu N_0 \frac{h^2}{\Omega r} & -\frac{1}{2}\mu d h r \\ -\frac{1}{2}\mu d h r & 0 \end{bmatrix}, \quad \mathbf{D}_{\text{pad}} = \begin{bmatrix} 2dr^2 & 0 \\ 0 & 0 \end{bmatrix},$$

$$\mathbf{D}_{\text{disc}}(t) = \begin{bmatrix} d_t(1 + \frac{1}{2}\delta) & 0 \\ 0 & d_t(1 + \frac{1}{2}\delta) \end{bmatrix} - \frac{1}{2}\delta d_t \begin{bmatrix} \cos(2\Omega t) & \sin(2\Omega t) \\ \sin(2\Omega t) & -\cos(2\Omega t) \end{bmatrix}, \qquad (3.3)$$

$$\mathbf{D}_{m_p}(t) = -\Theta_p \Omega \begin{bmatrix} -\sin(2\Omega t) & \cos(2\Omega t) \\ \cos(2\Omega t) & \sin(2\Omega t) \end{bmatrix}.$$

3.2 Optimization Technique

Introducing modified damping parameters $\tilde{d}_t = \alpha_1 d_t$ and $\tilde{d} = \alpha_2 d$ yields the vector $\hat{\mathbf{p}} = (\alpha_1, \alpha_2)^{\mathsf{T}}$ to be the set of parameters which is about to be optimized, cf. Sect. 2.5. Hence, damping due to the disc and damping due to the pins can be treated independently. Since the matrices (3.2) are time-dependent, FLOQUET theory is applied and the *monodromy matrix* $\mathcal{M}$ is calculated numerically. In order to make FLOQUET theory and CEA consistent, Eq. (2.22) is used to calculate the maximum real part of λ for a given μ. Setting $\delta = 0$, $\kappa = 0$, and $\Theta_p = 0$, both FLOQUET theory and CEA lead to the same results. Optimization problem (2.21) is sought using MATLAB, where an insight into the pseudocode is given below. To keep it short, the FFT and the optimization of the damping ratio are not included.

Pseudocode of optimization problem (2.21)

- for $\alpha_1 = \alpha_{1,\min}$ to $\alpha_{1,\max}$
- for $\alpha_2 = \alpha_{2,\min}$ to $\alpha_{2,\max}$

  – $\hat{\mathbf{p}} = (\alpha_1, \alpha_2)^{\mathsf{T}}$
  – calculate $\mathbf{\Gamma}\hat{\mathbf{p}}$
  – find $\tilde{\alpha}_1, \tilde{\alpha}_2$ satisfying $\mathbf{\Gamma}\tilde{\mathbf{p}} \le \mathbf{c}$, where $\tilde{\mathbf{p}} = (\tilde{\alpha}_1, \tilde{\alpha}_2)^{\mathsf{T}}$

- end
- end
- for each $\tilde{\mathbf{p}}$

  – $\tilde{d}_t = \tilde{\alpha}_1 d_t$
  – $\tilde{d} = \tilde{\alpha}_2 d$
  – calculate $\mathcal{M}$ using (2.18)
  – calculate $\mu_{\max} = \max\limits_{i} |\mu_i|$

- end
- find $\min_{\tilde{\mathbf{p}}} \mu_{\max}$ and output $\tilde{\alpha}_1 = \alpha_{1,\mathrm{opt}}$, $\tilde{\alpha}_2 = \alpha_{2,\mathrm{opt}}$
- calculate $\min_{\hat{\mathbf{p}}} \max_{i} \mathrm{Re}(\lambda_i)$ using (2.22)

As can be seen in the pseudocode, the *monodromy matrix* is calculated numerically for every vector $\tilde{\mathbf{p}}$ within the admissible area using (2.18), where the limit is set to $m = 100$. This way of optimization has an advantage and a disadvantage. On the one hand, calculating the *Floquet multipliers* respectively the eigenvalues for every $\tilde{\mathbf{p}}$ guarantees the calculated optimum point to be a global optimum. A discussion about local and global maxima is not necessary. On the other hand, this may lead to large computing times which are mainly influenced by the range of $\mathbf{p}$ and the increments. In this example, the increments are chosen as $\Delta\alpha_{1,2} = 0.01$ within a maximal range of $\alpha_{1,2} \in [0, 3]$ which may be interpreted as technically relevant. The degree of precision of the weighting factors then is ± 0.005.

3.3 Optimization Results

3.3.1 Time-Invariant Model

In this section, the system is assumed to be time-invariant, i.e. the parameters δ, κ, and θ_P are set to zero and only self-excitation due to circulatory terms may lead to unstable solutions, i.e. there is no parametric excitation. Applying the aforementioned optimization technique to determine the optimal weighting factors for the damping parameters d_t and d leads to the results in Fig. 3.2.

The dotted pattern ② represents the admissible area, where $\boldsymbol{\Gamma}\hat{\mathbf{p}} \leq \mathbf{c}$ is satisfied. Since the weighting factors can be varied independently, a rectangle is the simplest

Fig. 3.2 Optimization of d_t (weighting factor α_1) and d (weighting factor α_2), $\boldsymbol{\Gamma}$ and $\mathbf{c}$ are chosen as in (3.4). ①: flutter instability, admissible; ② \ ①: asymptotic stability, admissible; ③: asymptotic stability, inadmissible

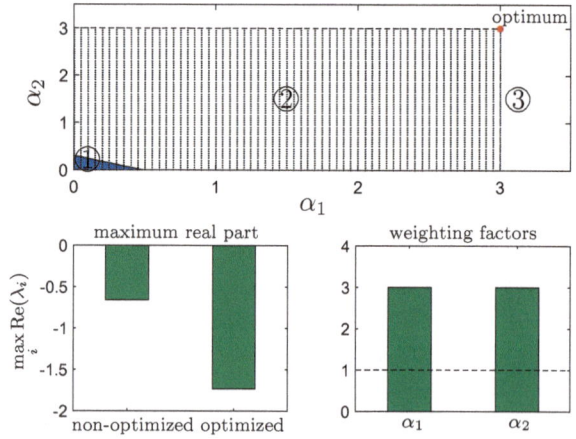

way of setting the constraints which are chosen as

$$\mathbf{\Gamma} = \begin{bmatrix} 1 & 0 \\ 0 & 1 \\ -1 & 0 \\ 0 & -1 \end{bmatrix}, \quad \mathbf{c} = \begin{bmatrix} 3 \\ 3 \\ 0 \\ 0 \end{bmatrix}. \tag{3.4}$$

The white area, $[③ \cup ②] \setminus ①$, including a subset of the admissible area, predicts the stable solution, whereas in the blue area $①$, the solution is unstable and of flutter-type. If only COULOMB damping is active, i.e. $\alpha_1 = \alpha_2 = 0$, the solution is unstable. The green bars compare the optimized and non-optimized maximum real part and the weighting factors, respectively. The red dot denotes the optimum $\tilde{\mathbf{p}}$. Since there is only one coherently blue area of flutter-type instability, which has the form of a triangle, neither adding damping in the disc nor in the pads, in this case, can destabilize the system. Larger values of α_1 and α_2 within the admissible area always stabilize or make more stable the equilibrium position. Therefore, the optimum is characterized by the largest possible values $\alpha_{1,\mathrm{opt}} = \alpha_{2,\mathrm{opt}} = 3$.

Thinking economically, an additional cost function may further reduce the scope of the admissible area. Similar to the operations research optimization, a cost function could have the form

$$c_1 \frac{\tilde{d}_t}{d_t} + c_2 \frac{\tilde{d}}{d} \leq c_{\max}, \tag{3.5}$$

where c_1 and c_2 represent the costs of adding damping, while the maximum costs, which may be due to additional economic aspects, are limited by $c_{\max}$. Setting for example $c_1 = 1000$, $c_2 = 1000$, and $c_{\max} = 3000$ yields $\alpha_1 + \alpha_2 \leq 3$. Due to physical aspects, the damping parameters have to be larger or equal than zero, i.e. $\alpha_1 \geq 0$, $\alpha_2 \geq 0$, which is equivalent to $-\alpha_1 \leq 0$ and $-\alpha_2 \leq 0$. Finally, the constraints can be written as

$$\mathbf{\Gamma} = \begin{bmatrix} 1 & 1 \\ -1 & 0 \\ 0 & -1 \end{bmatrix}, \quad \mathbf{c} = \begin{bmatrix} 3 \\ 0 \\ 0 \end{bmatrix}, \tag{3.6}$$

where the admissible area is a triangle that excludes the optimal point calculated above. In this case, the optimum point is $\alpha_{1,\mathrm{opt}} = 1.16$ and $\alpha_{2,\mathrm{opt}} = 1.84$, cf. Fig. 3.3. With regard to decomposition (3.3) and to the constraints (3.6), it therefore can be concluded that it is reasonable to shift some damping from the disc to the pins for system (2.5) with the matrices (3.2) and the parameters (3.1) to be optimized for stability.

In Fig. 3.4, a contour plot combined with a gradient field provides information on how to interpret these results. The contours represent the values of the maximum real part depending on the weighting factors α_1 and α_2, while the arrows represent the gradient of the maximum real part as a function of $\hat{\mathbf{p}} = (\alpha_1, \alpha_2)^{\mathrm{T}}$,

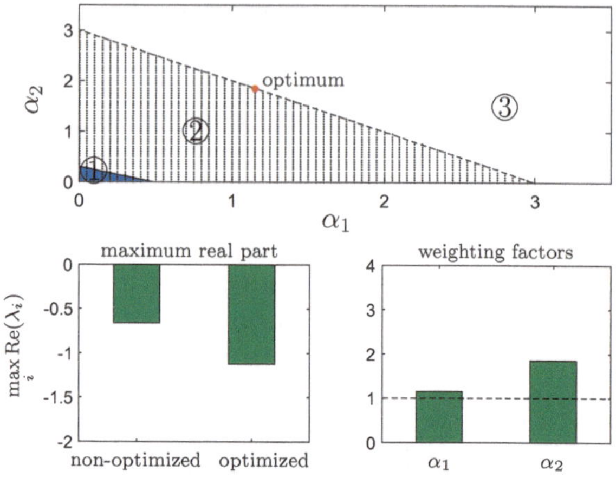

Fig. 3.3 Optimization of d_t (weighting factor α_1) and d (weighting factor α_2), $\boldsymbol{\Gamma}$ and $\mathbf{c}$ are chosen as in (3.6). ①: flutter instability, admissible; ② \ ①: asymptotic stability, admissible; ③: asymptotic stability, inadmissible

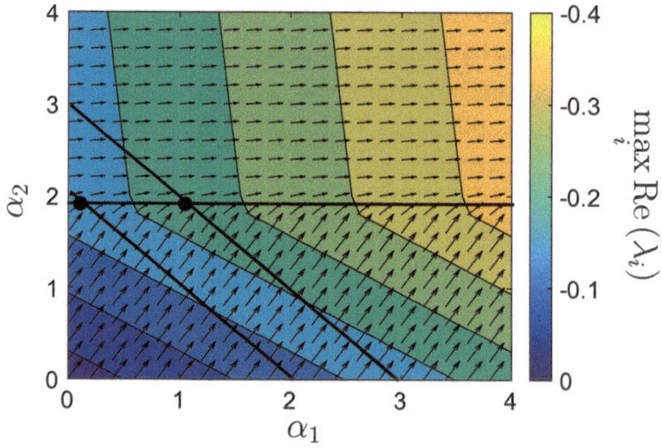

Fig. 3.4 Gradient field and contour plot of d_t (weighting factor α_1) and d (weighting factor α_2)

i.e. $\frac{\partial}{\partial \mathbf{p}}[\max_i \mathrm{Re}\,(\lambda_i)]$ and thus indicate the most beneficial direction to make more stable the equilibrium solution.

It becomes apparent that adding damping in the pads is only worthwhile up to $\alpha_{2,\mathrm{lim}} \approx 2$, whereas damping in the disc is always advantageous. Evaluating the gradient field with respect to the calculated optimum point in Fig. 3.2, i.e. $\alpha_{1,\mathrm{opt}} = \alpha_{2,\mathrm{opt}} = 3$, more damping in the pads than $\tilde{d}_t = \alpha_{2,\mathrm{lim}}d_t$ may not be beneficial. Choosing any $\alpha_2 > \alpha_{2,\mathrm{lim}}$ makes nearly no difference with regard to the maximum real part, whereas a larger value of α_1 is always purposeful. For this reason, damping in the disc can be given a higher priority. Evaluating Fig. 3.3 at $\alpha_{1,\mathrm{opt}} = 1.16$ and $\alpha_{2,\mathrm{opt}} = 1.84$ shows that the optimum value $\alpha_{2,\mathrm{opt}}$ is a more precise numerical calculation of $\alpha_{2,\mathrm{lim}}$, i.e. $\alpha_{2,\mathrm{opt}} = \alpha_{2,\mathrm{lim}} = 1.84$. For the constraints

$$\boldsymbol{\Gamma} = \begin{bmatrix} 1 & 1 \\ -1 & 0 \\ 0 & -1 \end{bmatrix}, \quad \mathbf{c} = \begin{bmatrix} c \\ 0 \\ 0 \end{bmatrix}, \tag{3.7}$$

with $c > 1.84$, the optimum value for damping in the pins is always $\alpha_{2,\mathrm{opt}} = 1.84$. Additionally, the condition $\alpha_1 + \alpha_2 = c$ has to be satisfied in order to minimize the maximum real part. Hence, the optimum value for damping in the disc can be determined by $\alpha_{1,\mathrm{opt}} = c - 1.84$. Setting

$$\mathbf{\Gamma} = \begin{bmatrix} 1 & 1 \\ -1 & 0 \\ 0 & -1 \end{bmatrix}, \quad \mathbf{c} = \begin{bmatrix} 2 \\ 0 \\ 0 \end{bmatrix} \tag{3.8}$$

yields $\alpha_{1,\mathrm{opt}} = 0.16$, which is the same result obtained in [6]. The points of intersection between the two skewed constraint lines and the horizontal line $\alpha_2 = 1.84$ in Fig. 3.4 illustrate the position of the two optimum points discussed above. It is important to mention that, in most cases, the optimum point is located at a corner of the admissible area, whereas in the example above the optimum is located at an edge. Certainly, these results are not based on analytical calculations and are to be understood as numerical approximations in a finite area. However, no counterexamples for these results in $(\alpha_1, \alpha_2) \in [0, 100] \times [0, 100]$ could be found in this study.

Since the damping ratio (2.20) is a favoured physical quantity to evaluate the rate of decay of damped oscillations, optimization (2.19) with $\max\limits_{\hat{\mathbf{p}}} \min\limits_{i} D_i(\lambda_i)$ is applied with the constraints defined by (3.6). As shown in Fig. 3.5, the optimum at $\alpha_{1,\mathrm{opt}} = 1.16$ and $\alpha_{2,\mathrm{opt}} = 1.84$ matches the optimum point in Fig. 3.3. This is due to the fact that the frequency of a damped mode is nearly the same as in its undamped equivalent. However, the difference between the damped and the undamped frequency may have an influence when optimizing the stiffness instead of the damping parameters. The magnitude of the damping ratio can be explained by its definition. Since the magnitude of the frequency is $\mathcal{O}(10^3)$ and the maximum real part is $\mathcal{O}(10^{-1})$, Eq. (2.20) results in a damping ratio being $\mathcal{O}(10^{-4})$.

A contour plot with the weighting factors being in the extended intervals $0 < \alpha_1 < 30$ and $0 < \alpha_2 < 1000$ shows that adding damping may also lead to larger maximum real parts of the eigenvalues, cf. Fig. 3.6. Every contour line has a turning point at $\alpha_2 \approx 120$, so damping in the pads in general may not make the system more stable. Numerical investigations using arbitrarily large weighting factors show, however, that adding damping in this brake model cannot destabilize the system. For $\alpha_2 > 120$ the maximum real part converges asymptotically to zero. Choosing $\alpha_1 = \alpha_1^*$ arbitrarily and optimizing the damping in the pads yields $\alpha_{2,\mathrm{opt}} \approx 120$ which can be assumed to be a global optimum in $\alpha_2 \in \mathbb{R}^+$. At this point, damping in the pads would have to be around 120 times higher than the initial value d given in (3.1) for which reason a technical relevance of this global optimum is questionable.

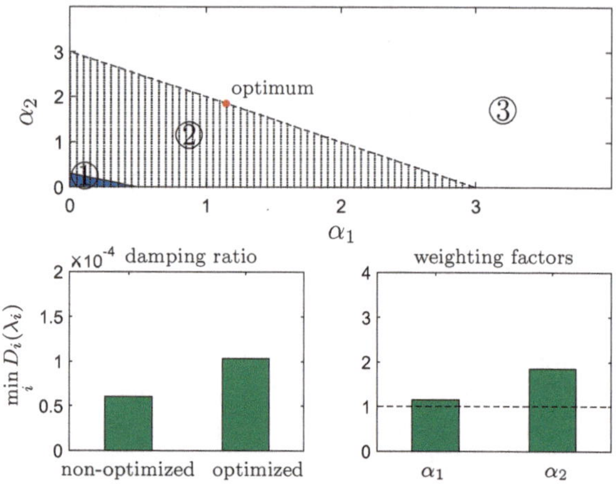

Fig. 3.5 Optimization of d_t (weighting factor α_1) and d (weighting factor α_2), Γ and $\mathbf{c}$ are chosen as in (3.6). ①: flutter instability, admissible; ② \ ①: asymptotic stability, admissible; ③: asymptotic stability, inadmissible

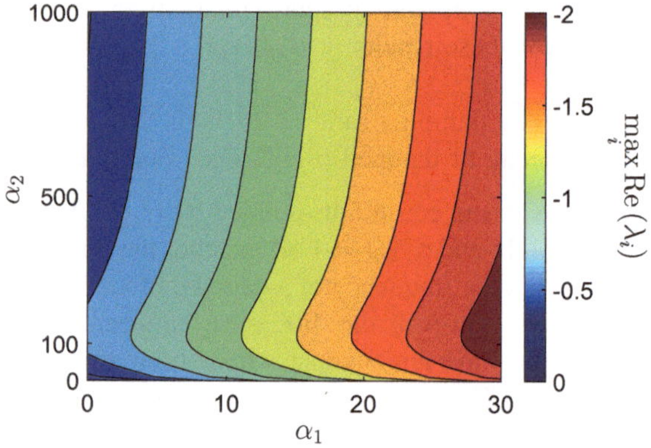

Fig. 3.6 Contour plot of d_t (weighting factor α_1) and d (weighting factor α_2)

3.3.2 Time-Periodic Model

In this section, damping of the brake model is optimized for the more general case of time-periodic matrices. As can be seen in (3.2), the parameters δ, κ, and Θ_p activate time-variance and may force the system to parametrically excited vibrations. In comparison to the case of symmetric damping in the bearing, numerical investigations suggest that asymmetric damping ($\delta \neq 0$) nearly has no influence on the stability. Thus, only the perturbation parameters $\kappa \neq 0$ and $m_p \neq 0$ are discussed which are chosen to compare with [7], i.e.

$$\delta = 0, \ \kappa = 0.02, \ \Theta_p = -0.001. \tag{3.9}$$

A negative moment of inertia respectively a particle with negative mass can be interpreted as a hole in the disc which, in realistic brakes, has the purpose of draining water off to improve the braking properties. Consider the hole having the distance $r_p = 0.1$m to the center of the disc and the radius $r_h = 0.0025$m. The disc is made out of steel (density $\rho_{steel} \approx 7.9$ g/cm^3) and is $h = 0.02$m thick. Since $r_h \ll r_p$, the approximation of the hole to be a mass point can be used. Calculating the moment of inertia $\Theta_p = \frac{1}{2}\rho_{steel}\pi r_h^2 h r_p^2$ yields $\Theta_p = \mathcal{O}(10^{-3})$.

With the linear constraints $\mathbf{\Gamma}$ and $\mathbf{c}$ as in (3.4) the optimum still lies in the upper right corner of the admissible range, where $\alpha_{1,\text{opt}} = \alpha_{2,\text{opt}} = 3$. Compared to the time-invariant case, the region of instability (blue triangle) grows. However, setting κ sufficiently large, e.g. $\kappa = 0.07$, this domain disappears completely. Using constraints (3.6) results in a different optimum point which is at $\alpha_{1,\text{opt}} = 0$ and $\alpha_{2,\text{opt}} = 3$, cf. Fig. 3.7. The course of the contour lines and the direction of the arrows in Fig. 3.8 are similar compared to Fig. 3.4. The line $\alpha_2 = \alpha_{2,\text{lim}}$, where increasing α_2, i.e. adding damping in the pads, starts being no longer worthwhile, moves upwards from $\alpha_2 \approx 1.84$ to $\alpha_2 \approx 6.3$. Hence, adding damping in the pads is more efficient than in the time-invariant case. The optimization of the damping ratio according to (2.24) yields the same results.

Fig. 3.7 Optimization of d_t (weighting factor α_1) and d (weighting factor α_2), $\mathbf{\Gamma}$ and $\mathbf{c}$ are chosen as in (3.6). ①: flutter instability, admissible; ② \ ①: asymptotic stability, admissible; ③: asymptotic stability, inadmissible

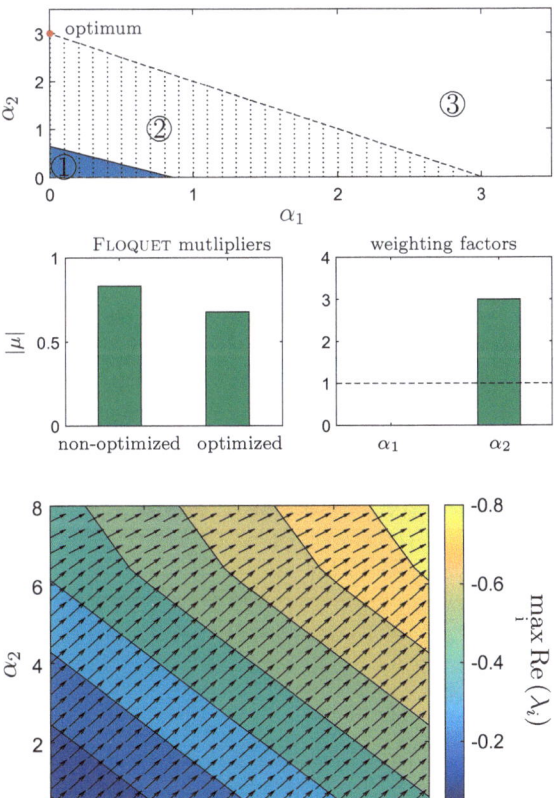

Fig. 3.8 Gradient field and contour plot of d_t (weighting factor α_1) and d (weighting factor α_2), where the maximum real part is calculated with (2.23)

3.3.3 Discussion

It is well known from practical experience that brake squeal depends on various parameters, especially on the prestress N_0 and the friction coefficient μ. However, numerical investigations in this study show, that stability maps and contour plots are geometrically similar when varying these parameters. For example, setting N_0 small, the real parts of the eigenvalues and hence the blue triangle, which represents the area of flutter-type instability, grow and vice versa. Setting N_0 as small as possible within the velocity range where brake squeal occurs, would be a possibility to guarantee optimum damping parameters. The prediction of the model that large prestress forces act stabilizing, whereas small prestress forces act destabilizing matches with reality since the noise phenomenon mainly occurs at low vehicle speeds when the driver applies the brake gently.

Another key value is the angular velocity Ω. The maximum real part of λ grows with Ω such that there exists a limit angular velocity above which the system is unstable [1]. This result seems to be a contradiction to practical experience. As shown in [3], the friction coefficient μ depends on Ω for high velocity values. Setting μ as a linear decreasing function of the angular speed, i.e. $\mu(\Omega) = \mu_0 - m\Omega$, where $m > 0$, a decreasing maximum real part for growing Ω can be observed. In this study, however, only small angular velocities are investigated and the restriction of a constant friction coefficient remains reasonable.

3.4 Traps and Shortcomings of CEA

With respect to time-periodic systems, it should be noted that using CEA, as it is normally done in the automotive industry, leads into a trap. In [7], it is shown that the eigenvalues of a time-dependent matrix $\mathbf{A}(t)$, calculated at some frozen times over one period, provide no information about the stability. The system may be asymptotically stable although there exist eigenvalues with positive real part. Similarly, the system may be unstable although all eigenvalues calculated at discrete time steps have negative real part. Let $\lambda(t)$ be the solution of the equation $\det(\mathbf{A}(t) - \lambda(t)\mathbf{I}) = 0$. Then, it is not possible to draw conclusions on the *Floquet multipliers* μ from $\lambda(t)$, i.e.

$$\exists\, \tilde{t} \in [0, T]\, \exists\, i \in [0, 2n] : \mathrm{Re}(\lambda_i(\tilde{t})) > 0 \nRightarrow \exists\, \mu_j : |\mu_j| > 1, \qquad (3.10)$$

respectively

$$\forall\, \tilde{t} \in [0, T]\, \forall\, i \in [0, 2n] : \mathrm{Re}(\lambda_i(\tilde{t})) < 0 \nRightarrow \forall\, \mu_j : |\mu_j| < 1. \qquad (3.11)$$

Conclusion (3.10) is derived using a time-periodic system with one DOF. The two complex conjugated eigenvalues can be represented by $\lambda_1(t) = a(t) + ib(t)$ and

$\lambda_2(t) = a(t) - ib(t)$. The LIOUVILLE formula is

$$\det[\boldsymbol{\Phi}(t)] = \det[\boldsymbol{\Phi}(0)]e^{\int_0^t \text{tr}[\mathbf{A}(\tau)]d\phi}. \tag{3.12}$$

Without loss of generality, the initial conditions may be set equal to the identity matrix and the time interval of the integral in (3.12) is chosen as one period T. Furthermore, the trace and the determinant of a matrix are known to be the sum and the product of its eigenvalues, respectively. As a result, Eq. (3.12) simplifies to

$$e^{2\int_0^T a(t)dt} = \mu_1\mu_2. \tag{3.13}$$

Using the mean value theorem of integration $\int_0^T a(t)dt = a(\xi)T$ for $\xi \in [0, T]$ Eq. (3.13) can be rewritten as

$$e^{2a(\xi)T} = \mu_1\mu_2. \tag{3.14}$$

It is possible that the mean value $a(\xi)$ is negative, while there exists any $a(\tilde{t}) > 0$. Adopting this situation yields

$$\mu_1\mu_2 < 1. \tag{3.15}$$

If the *Floquet multipliers* are assumed to be complex, i.e. $\mu_1\mu_2 = |\mu_1|^2 = |\mu_2|^2$, it follows

$$|\mu_1| = |\mu_2| < 1. \tag{3.16}$$

Therefore, the system may be asymptotically stable although there are positive real parts at all times during one period T.

Expanding (3.12) to an n-dimensional problem leads to similar results. In this more general case, the LIOUVILLE formula becomes

$$e^{\int_0^T \sum_{i=1}^n \lambda_i(t)dt} = \prod_{i=1}^n \mu_i. \tag{3.17}$$

Suppose the eigenvalue λ_n to be larger than zero during one period, i.e. $\lambda_n(t) > 0 \ \forall \ t \in [0, T]$. Splitting λ_n from the integral Eq. (3.17) becomes

$$e^{\int_0^T \sum_{i=1}^{n-1} \lambda_i(t)dt + \int_0^T \lambda_n(t)dt} = \prod_{i=1}^n \mu_i. \tag{3.18}$$

If the condition

$$\int_0^T \sum_{i=1}^{n-1} \lambda_i(t)dt + \int_0^T \lambda_n(t)dt < 0 \tag{3.19}$$

is satisfied, Eq. (3.18) becomes

$$\prod_{i=1}^{n} \mu_i < 1. \tag{3.20}$$

On the one hand, the product of n *Floquet multipliers* being smaller than one may contain factor values larger than one or only *Floquet multipliers* smaller than one and no conclusion about the largest *Floquet multiplier* can be drawn. On the other hand, consider the case

$$\int_0^T \sum_{i=1}^{n} \lambda_i(t)dt = 0. \tag{3.21}$$

Then, Eq. (3.18) reads

$$\prod_{i=1}^{n} \mu_i = 1 \tag{3.22}$$

and it can be concluded that either there exists at least one *Floquet multiplier* with an absolute value larger than one or each *Floquet multiplier* has an absolute value equal to one. Hence, the system either is weakly stable or unstable. Consequently, if

$$\int_0^T \sum_{i=1}^{n} \lambda_i(t)dt > 0 \tag{3.23}$$

is satisfied, the system must be unstable. As a result, condition (3.19) is necessary for a time-periodic system to be asymptotically stable. In most cases, this condition is fulfilled for damped systems since it is equivalent to

$$-\int_0^T \text{tr}[\mathbf{A}(t)]dt = \int_0^T \text{tr}[\mathbf{M}^{-1}(t)\mathbf{D}(t)]dt > 0 \tag{3.24}$$

which is satisfied if $\mathbf{D}(t) > 0 \; \forall \, t \in [0, T]$. Furthermore, since $\int_0^T \text{tr}[\mathbf{A}(t)]dt = 0$, if $\mathbf{D}(t) = 0 \; \forall t \in [0, T]$, it can be seen with regard to condition (3.21), that an undamped time-periodic system, similar to the case of time-invariant systems, cannot be asymptotically stable; it may only be weakly stable.

The error that occurs using CEA for some frozen times in technically time-periodic models is shown in Fig. 3.9. The parameter β_1 on the abscissa is the weighting factor associated with the stiffness k_t of the disc and the parameter β_2 on the ordinate is the weighting factor associated with the stiffness k of the pins, both of which are are chosen as in (3.1). The comparison between the stability maps computed using FLOQUET theory and those applying the concept of frozen times demonstrate the trap. The former predicts circle-like areas of instability, whereas the latter yields misleading results. At certain times, e.g. at $t = 0$, a fan-shaped instability area appears. Within the time interval $\frac{2}{10}T < t < \frac{8}{10}T$ this area totally disappears and reappears for $t \in (\frac{8}{10}T, T]$.

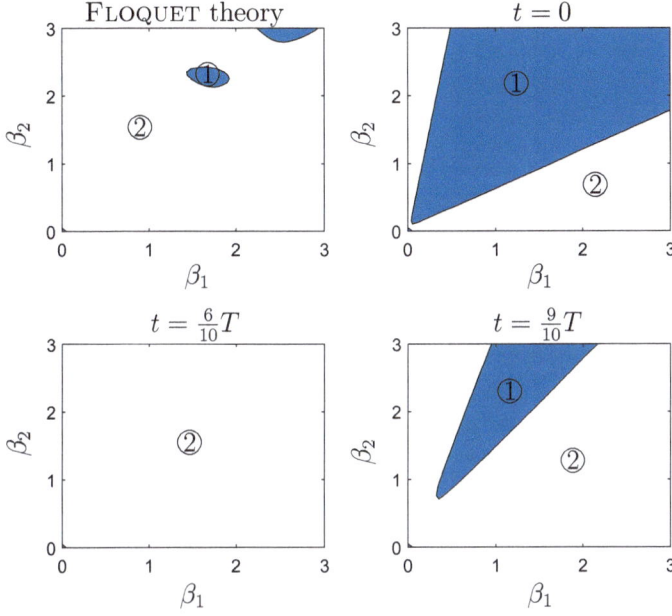

Fig. 3.9 Stability maps of the parameters k_t (weighting factor β_1) and k (weighting factor β_2); top left: Stability map using FLOQUET theory; others: Stability maps for several frozen times; ①: flutter instability, ②: asymptotic stability

Since (3.17) is equivalent to

$$e^{\int_0^T \mathrm{tr}[-\mathbf{M}^{-1}(t)\mathbf{D}(t)]\mathrm{d}t} = \prod_{i=1}^{n} \mu_i, \qquad (3.25)$$

it can be concluded that the structure of a time-dependent damping matrix plays also an important role with regard to stability. Consider the equations of motion of the wobbling disc with the matrices (3.2) and $m_p = 0$. To ensure the system to be either unstable or weakly stable the damping matrix $\mathbf{D}$ must satisfy

$$\mathrm{tr}[\mathbf{M}^{-1}\mathbf{D}(t)] = 0. \qquad (3.26)$$

Introducing the mass and damping matrix from (3.2) yields

$$\delta = -2\frac{dr^2}{d_t} - \frac{\mu N_0 h^2}{2d_t \Omega r} - 2. \qquad (3.27)$$

As discussed in Sect. 3.3.2, numerical investigations suggest that asymmetric damping properties only have little influence on the stability behavior of the investigated brake model. Especially, a destabilization of the system, when $\delta \neq 0$, is not possible because $\delta > 0$ and (3.27) cannot be satisfied simultaneously. However, since the system may be unstable (or stable), if $\int_0^T \mathrm{tr}[\mathbf{M}^{-1}\mathbf{D}(t)]\mathrm{d}t < 0$, there may be unknown values δ which have a more significant influence on the system's stability.

References

1. von Wagner, U., Hochlenert, D., Hagedorn, P.: Minimal Models for Disk Brake Squeal. J. Sound Vib. **302**(3), 527–539 (2007)
2. Hochlenert, D.: Selbsterregte Schwingungen in Scheibenbremsen: Mathematische Modellbildung und aktive Unterdrückung von Bremsenquietschen. Berichte aus dem Maschinenbau. Shaker, Aachen (2006)
3. Breuer, B., Bill, K.H. (eds.): Bremsenhandbuch: Grundlagen, Komponenten, Systeme, Fahrdynamik. ATZ/MTZ-Fachbuch, 4th edn. Springer Vieweg, Wiesbaden (2013)
4. Spelsberg-Korspeter, G., Hochlenert, D., Hagedorn, P.: Non-linear investigation of an asymmetric disk brake model. Proc. IMechE Part C: J. Mech. Eng. Sci. **225**(10), 2325–2332 (2011)
5. Hagedorn, P., Eckstein, M., Heffel, E., Wagner, A.: Self-excited vibrations and damping in circulatory systems. J. Appl. Mech. **81**(10):101009–1–9 (2014)
6. Jekel, D., Clerkin, E., Hagedorn, P.: Robust damping in self-excited mechanical systems. Proc. Appl. Math. Mech. **16**(1), 695–696 (2016)
7. Spelsberg-Korspeter, G.: Eigenvalue optimization against brake squeal: symmetry, mathematical background and experiments. J. Sound Vib. **331**(19), 4259–4268 (2012)

Chapter 4
Optimization of Finite Element Models of Disc Brakes

4.1 Theoretical Background

The assumptions made to derive the equations of motion in a FE environment are similar to those of the minimal model. Since the FE method works with non-differentiable but continuous functions, it is assumed that there is no sticking and the disc rotates constantly such that COULOMB's law can be used for the contact analysis between the brake disc and the brake pads. A constant prestress force is loaded on the pads leading to additional velocity and coordinate proportional matrices with respect to an inertial coordinate system. The linearized equations of motion in PERMAS notation [1] read

$$\mathbf{M\ddot{q}} + (\mathbf{D}_V + \mathbf{D}_G + \mathbf{D}_F)\mathbf{\dot{q}} + (\mathbf{K} + \mathbf{K}_G + \mathbf{K}_C + \mathbf{K}_F + i\mathbf{H})\mathbf{q} = \mathbf{0}, \qquad (4.1)$$

where the physical meaning of the respective matrices is:

$\mathbf{M}$ mass matrix
$\mathbf{D}_V$ viscous damping matrix
$\mathbf{D}_G$ gyroscopic matrix
$\mathbf{D}_F$ friction induced damping matrix
$\mathbf{K}$ elastic stiffness matrix
$\mathbf{K}_G$ geometric stiffness matrix
$\mathbf{K}_C$ convective stiffness matrix
$\mathbf{K}_F$ correction matrix of contact forces
$\mathbf{H}$ structural damping matrix

For rotordynamical systems it is often required to generate a CAMPBELL diagram which represents a system's response spectrum as a function of its oscillation regime. In PERMAS, the angular velocity of the brake disc can be varied a posteriori, i.e. after the equations of motion (4.1) have been derived. The circular frequency ω_j scales the corresponding matrices with respect to a reference circulatory

© The Authors 2018
J.-H. Wehner et al., *Damping Optimization in Simplified and Realistic Disc Brakes*, SpringerBriefs in Applied Sciences and Technology,
DOI 10.1007/978-3-319-62713-7_4

Table 4.1 Velocity proportional output matrices from PERMAS

Output name	Physical meaning	Scaling factor
BDMRLL	Viscous damping including velocity dependent damping for $\omega_j = \omega_{\mathrm{ref}}$	1
BDIWMLL	Velocity dependent damping (friction induced)	$\frac{\omega_{\mathrm{ref}}}{\omega_j} - 1$
BYMLL	Gyroscopic terms	$\frac{\omega_j}{\omega_{\mathrm{ref}}}$
BHMRLL	Structural damping	$\frac{1}{2\pi f_{\mathrm{cev}}}$

frequency ω_{ref}. In Table 4.1, the velocity proportional output matrices from PERMAS are listed including their physical meaning and the corresponding scaling factor.

Combining the output matrices in Table 4.1 and using (2.36) for the calculation of the equivalent viscous damping matrix $\mathbf{D}_S$ yields the velocity proportional matrices

$$\mathbf{D}_V = \mathrm{BDMRLL} - \mathrm{BDIWMLL}, \qquad \mathbf{D}_G = \frac{\omega_j}{\omega_{\mathrm{ref}}} \mathrm{BYMLL},$$

$$\mathbf{D}_F = \frac{\omega_{\mathrm{ref}}}{\omega_j} \mathrm{BDIWMLL}, \qquad \mathbf{D}_S = \frac{1}{2\pi f_{\mathrm{cev}}} \mathrm{BHMRLL}. \tag{4.2}$$

Brake squeal usually occurs at a specific frequency range (mainly between 1 and 12 kHz). In order to reduce the number of DOF it is advantageous to transform the matrices into a modal subspace only containing the first k modes with their eigenfrequencies being in the frequency interval to be optimized. The upper frequency limit f_{lim} determines the number of modes being of interest. However, there is no equation relating f_{lim} and k. For a given upper frequency limit PERMAS calculates automatically the corresponding k eigenvalues [2].

The first step of a dynamic analysis usually is to determine the natural frequencies and mode shapes with damping neglected [3]. In PERMAS, a real eigenvalue analysis establishes the basis for the modal transformation. Assuming the mass and the stiffness matrix to be positive definite, the eigenvalue analysis of the corresponding **MK**-system yields the undamped, natural frequencies and guarantees real eigenvectors representing the columns of the modal matrix. Since damping only has little influence on the natural eigenfrequencies and eigenvectors, these results characterize the basic dynamic behavior of the structure. The modal matrix of the **MK**-system can then be used to transform the general **MDGKN**-system from the physical space into the modal subspace. The modal transformation of the minimal coordinates reads

$$\mathbf{q} = \mathbf{X}_e \mathbf{p}, \tag{4.3}$$

where $\mathbf{X}_e$ is the modal matrix containing the eigenvectors of the elastic mode shapes of the **MK**-system. Modal reduction now means to approximate the vector of the generalized coordinates $\mathbf{q}$ by $m \ll n$ eigenvectors, i.e.

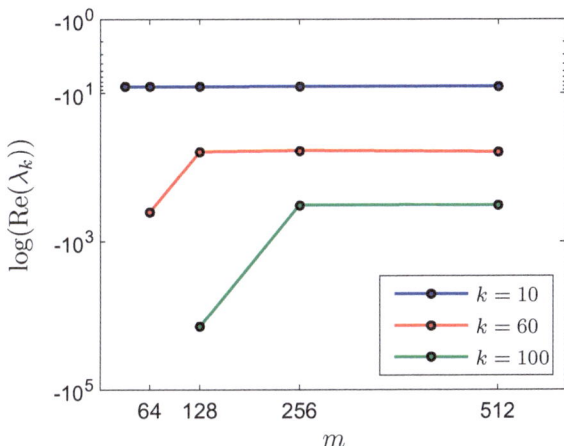

Fig. 4.1 Convergence analysis of the model described in Sect. 4.2, where m is associated with the number of calculated eigenvalues and k represents the k-th eigenvalue

$$\mathbf{q} \approx \mathbf{X}_{e,m}\mathbf{p}, \tag{4.4}$$

and to transform the equations of motion (4.1) according to (2.39), where m is determined by the frequency range of interest. Modal reduction avoids calculating irrelevant eigenvalues and the computing time of CEA can be reduced drastically. The technique guarantees that only those modes being of interest are to be optimized.

In Fig. 4.1, a convergence analysis of the simplified FE model of disc brake described in Sect. 4.2 is conducted. The number of calculated eigenvalues m is given on the abscissa, whereas on the ordinate the logarithmized real part of the k-th eigenvalue is plotted. The colored lines represent the dependency of the real part of the k-th eigenvalue on the number m of calculated eigenvalues, i.e. a function $f(m) = \text{Re}(\lambda_k)(m)$. The cases $k = 10$, $k = 60$, and $k = 100$, which are to be understood as examples, show that more eigenvalues than being of interest have to be calculated in order to guarantee the k-th eigenvalue to be converged. Analyzing the 10-th eigenvalue and calculating $m = 64$ eigenvalues is no problem with respect to convergence. In the case $k = 100$, however, it is necessary to calculate at least about 2.5 times more eigenvalues. Of course, studying the convergence of the imaginary parts respectively the eigenfrequencies leads to similar results.

For practical evaluations in MATLAB, the vector $\mathbf{E}_m$ containing $m \geq 2.5k$ eigenvalues from the reduced m-DOF system is calculated and the k eigenvalues being of interest in the vector $\mathbf{E_k}$ are cut off afterwards, i.e.

$$\mathbf{E}_m = \begin{bmatrix} \lambda_1 \\ \lambda_2 \\ \vdots \\ \lambda_k \\ \lambda_{k+1} \\ \vdots \\ \vdots \\ \lambda_m \end{bmatrix} \left. \begin{matrix} \\ \\ \\ \\ \end{matrix} \right\} k \text{ eigenvalues} \qquad \left. \begin{matrix} \\ \\ \\ \\ \\ \\ \\ \\ \\ \end{matrix} \right\} \begin{matrix} m \geq 2.5k \\ \text{eigenvalues} \end{matrix} \qquad \mathbf{E}_k = \begin{bmatrix} \lambda_1 \\ \lambda_2 \\ \vdots \\ \lambda_k \end{bmatrix}.$$

Consequently, when optimizing the damping properties of the brake model, only the vector $\mathbf{E}_k$ containing the first k eigenvalues has to be considered.

Although the damping properties in this study are optimized in a modal subspace, it is possible to transfer back the results to the physical space. The weighting factors as a set of parameters $\hat{\mathbf{p}} = (\alpha_1, \alpha_2, ..., \alpha_n)^{\mathrm{T}}$ are multiplied with different modal damping matrices

$$\tilde{\mathbf{D}} = \sum_{i=1}^{n} \alpha_i \tilde{\mathbf{D}}_i. \tag{4.5}$$

The back transformation from modal into physical coordinates then reads

$$\mathbf{D} = \mathbf{X}_{e,m} \tilde{\mathbf{D}} \mathbf{X}_{e,m}^{\mathrm{T}} \tag{4.6}$$

and hence

$$\mathbf{D} = \sum_{i=1}^{n} \mathbf{X}_{e,m} \alpha_i \tilde{\mathbf{D}}_i \mathbf{X}_{e,m}^{\mathrm{T}}, \tag{4.7}$$

where $\mathbf{X}_{e,m} \in \mathbb{R}^{n \times m}$. Since scalars are invariant with regard to coordinate-transformations, the damping matrix in physical coordinates is

$$\mathbf{D} = \sum_{i=1}^{n} \alpha_i \mathbf{D}_i, \tag{4.8}$$

where each damping matrix $\mathbf{D}_i$ has a distinct physical origin, e.g. $\mathbf{D}_V$ (viscous damping), $\mathbf{D}_F$ (friction-induced damping), or $\mathbf{D}_S$ (structural damping). With this method, the weighting factors scale the damping matrix in physical coordinates respectively the real part of the eigenvalues within the frequency range of interest. A disadvantage is, however, that optimizing the entries of the damping matrix independently is not possible. Firstly, optimizing each entry would drastically extend the computing time. Secondly, changing entries in the damping matrix individually means multiplying the damping matrix not by scalars but by weighting vectors or weighting matrices. A back transformation in the physical space may then cause problems because vectors and matrices are not invariant with regard to coordinate transformations.

The vector $\hat{\mathbf{p}} = (\alpha_1, \alpha_2)^{\mathsf{T}}$ containing two weighting factors represents the set of parameters which are about to be optimized. For example, studying viscous and structural damping yields a modified version of the equations of motion (4.1), i.e.

$$\mathbf{M}\ddot{\mathbf{q}} + (\alpha_1 \mathbf{D}_V + \alpha_2 \mathbf{D}_S + \mathbf{D}_G + \mathbf{D}_F)\dot{\mathbf{q}} + (\mathbf{K} + \mathbf{K}_G + \mathbf{K}_C + \mathbf{K}_F)\mathbf{q} = \mathbf{0}, \qquad (4.9)$$

the eigenvalues of which, respectively their real parts, can be optimized applying optimization (2.19). Multiplying the structural damping matrix $\mathbf{D}_S = \frac{g}{\Omega_{\mathrm{ref}}}\mathbf{K}$ is equivalent to multiplying the structural damping factor g by the weighting factor α_2 and allows introducing $\tilde{g} = \alpha_2 g$ as a modified parameter. Since the equations of motion (4.9) are time-invariant, a calculation of the *monodromy matrix* is not necessary. A short pseudocode of the optimization process is outlined below.

Pseudocode of optimization problem in modal space

- set $k = k^* \leq \frac{m}{2.5}$
- load the matrices of the modal space calculated in PERMAS
- for $\alpha_1 = \alpha_{1,\mathrm{min}}$ to $\alpha_{1,\mathrm{max}}$
- for $\alpha_2 = \alpha_{2,\mathrm{min}}$ to $\alpha_{2,\mathrm{max}}$

  – $\hat{\mathbf{p}} = (\alpha_1, \alpha_2)^T$
  – calculate $\boldsymbol{\Gamma}\hat{\mathbf{p}}$
  – find $\tilde{\alpha}_1, \tilde{\alpha}_2$ satisfying $\boldsymbol{\Gamma}\hat{\mathbf{p}} \leq \mathbf{c}$, where $\tilde{\mathbf{p}} = (\tilde{\alpha}_1, \tilde{\alpha}_2)^T$

- end
- end
- $\omega_{\mathrm{ref}} = 1$
- set $\omega_j = \omega_j^*$
- set $f_{\mathrm{cev}} = f_{\mathrm{cev}}^*$
- scale the damping matrices as described in Table 4.1 and scale the stiffness matrices
- for each $\tilde{\mathbf{p}}$

  – $\tilde{\mathbf{D}}_V = \tilde{\alpha}_1 \tilde{\mathbf{D}}_V$
  – $\tilde{g} = \tilde{\alpha}_2 g$
  – calculate $\mathbf{E}_m$
  – calculate $\mathbf{E}_{k^*}$
  – calculate $\lambda_{\mathrm{max}} = \max_{k^*} \mathrm{Re}\,(\lambda_{k^*})$

- end
- find $\min_{\tilde{\mathbf{p}}} \lambda_{\mathrm{max}}$ and output $\tilde{\alpha}_1 = \alpha_{1,\mathrm{opt}}$, $\tilde{\alpha}_2 = \alpha_{2,\mathrm{opt}}$

Fig. 4.2 Simplified brake
model from [4]

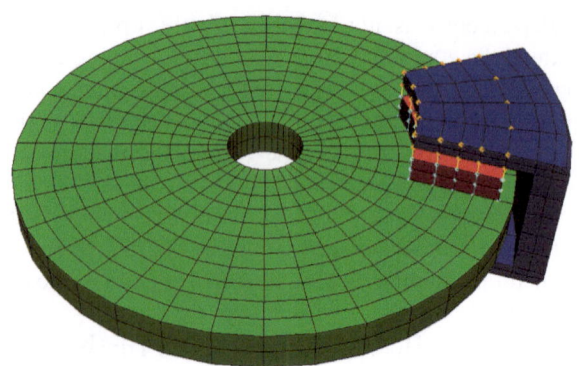

4.2 Low-Degree-of-Freedom Model

In this section, a simplified FE model of disc brake from [4] consisting of 4600 DOF
is used to explain the optimization technique in detail. As can be seen in Fig. 4.2, the
model consists of a disc rotating with constant angular speed, two pads in frictional
contact with the disc, and a caliper which is supported by springs against ground.
The pads, which are loaded by a constant pressure $p = 0.1$ MPa, are guided by rigid
elements fixed to the caliper. Additionally, they are supported by springs against it.
For the contact between the disc and the pads a constant sliding friction coefficient $\mu = 0.7$ is assumed. The reference angular velocity of the rotating disc is
$\omega_{\text{ref}} = 1\,\text{s}^{-1}$. Besides COULOMB and viscous damping, material damping in the
pads with a loss factor $g = 0.08$ is included.

4.2.1 Optimization Results

The application of optimization procedure (2.19) to the equations of motion (4.9)
in order to determine optimal weighting factors for the damping matrices $\mathbf{D}_V$ (viscous damping) and $\mathbf{D}_S$ (structural damping) leads to the result in Fig. 4.3. In this
example, the modal subspace was calculated with $k = 16$ and $m = 64$, where the
upper frequency limit is approximately 1 kHz. The angular speed of the disc is set
to $\omega_j = 2\pi\text{s}^{-1}$ and the linear constraints $\mathbf{\Gamma}$ and $\mathbf{c}$ are chosen according to (3.4). It
becomes apparent that these results are similar to those obtained in the investigation
of the minimal model in Sect. 3.3.1. On the one hand, the optimized maximum real
part is characterized by the largest values of the weighting factors within the admis-
sible area, i.e. $\alpha_{1,\text{opt}} = \alpha_{2,\text{opt}} = 3$. On the other hand, the blue area of flutter-type
instability still has approximately the shape of a triangle. Hence, adding damping
within the admissible area cannot act destabilizing; it always makes the system more
stable.

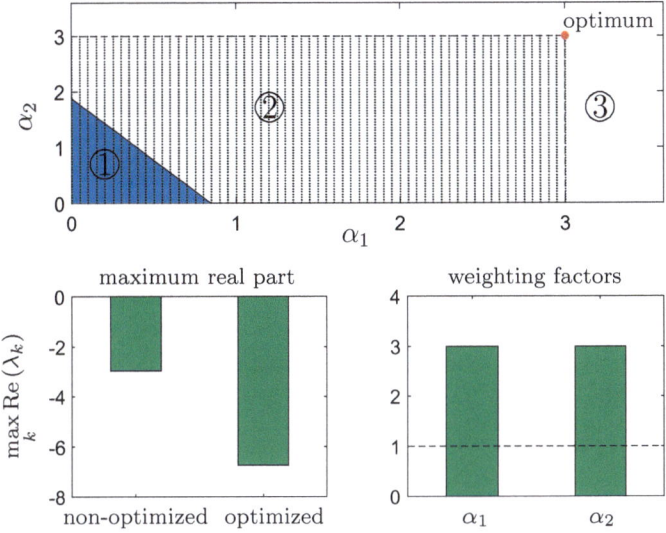

Fig. 4.3 Optimization of $\mathbf{D}_V$ (weighting factor α_1) and $\mathbf{D}_S$ (weighting factor α_2); $k = 16$, $m = 64$, $f_{\text{lim}} \approx 1\,\text{kHz}$. $\boldsymbol{\Gamma}$ and $\mathbf{c}$ are chosen as in (3.4). ①: flutter instability, admissible; ② \ ①: asymptotic stability, admissible; ③: asymptotic stability, inadmissible

Setting the upper frequency limit with approximately 3 kHz to a more reason-able value with regard to brake squeal yields a modal subspace with $k = 32$ and $m \geq 80$. The results can be seen in Fig. 4.4, where an increasing area of flutter-type instability can be observed and the non-optimized maximum real part is unstable. The magnitude of the non-optimized maximum real part grows about one magnitude compared to its equivalent in Fig. 4.3, so that the stabilization of a system with respect to higher frequencies requires more damping. Expanding the admissible area shows that both stability maps are geometrically similar. A growing unstable region can also be observed by setting higher values for the angular velocity ω_j. The same phenomenon, i.e. that a change of parameters leads to increasing or decreasing sta-bility maps with similar geometrical shapes, is already known from examining the minimal model.

The gradient field in combination with the contour plot in Fig. 4.5 helps to evalu-ate the results of Fig. 4.4. The different colors represent the values of the maximum real part in dependence of α_1 and α_2, while the direction and the length of the arrows represent the gradient of the maximum real part. Similar to the minimal model, there are values for the weighting factors beyond which adding more damping is no longer worthwhile, i.e. the maximum real part nearly stops decreasing. Here, such a limit line might be the dotted line in Fig. 4.5. However, this line has no exact mathemat-ical definition and can only be understood as an estimate. Furthermore, since the arrows become very small (they may even look like dots) when passing this line, this effect is greater compared to the minimal model. Nevertheless, the directions of the arrows are identical and the qualitative arguments which can be concluded

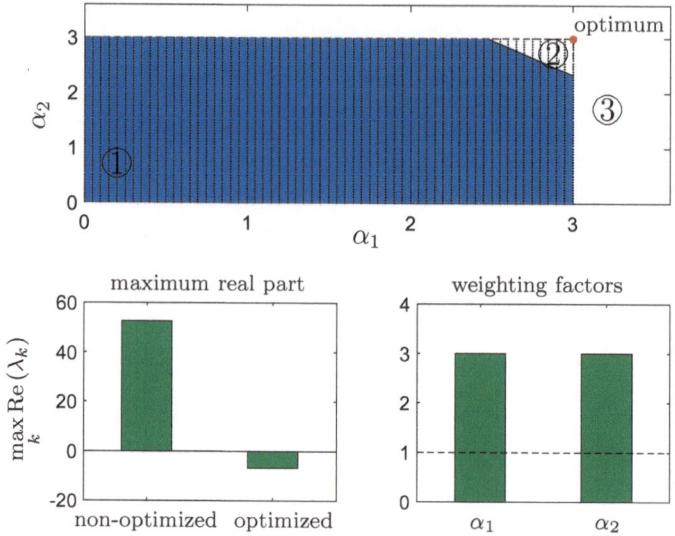

Fig. 4.4 Optimization of $\mathbf{D}_V$ (weighting factor α_1) and $\mathbf{D}_S$ (weighting factor α_2); $k = 32$, $m = 128$, $f_{\text{lim}} \approx 3\,\text{kHz}$. $\boldsymbol{\Gamma}$ and $\mathbf{c}$ are chosen as in (3.4). ①: flutter instability, admissible; ② \ ①: asymptotic stability, admissible; ③: asymptotic stability, inadmissible

Fig. 4.5 Gradient field and contour plot of $\mathbf{D}_V$ (weighting factor α_1) and $\mathbf{D}_S$ (weighting factor α_2); $k = 32$, $m = 128$, $f_{\text{lim}} \approx 3\,\text{kHz}$

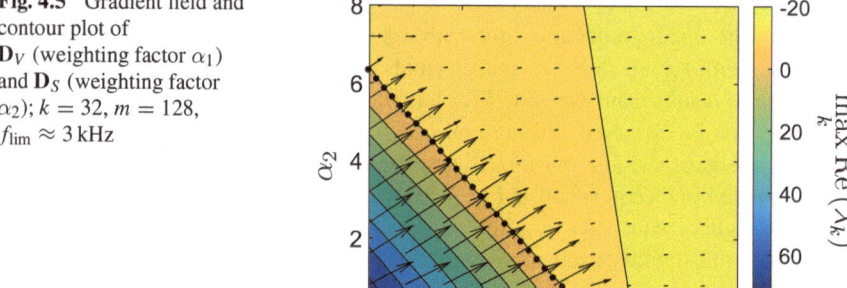

from both plots are similar. On the one hand, optimizing within a rectangular or quadratic admissible area containing weighting factors smaller than ten, which may be interpreted as technically relevant, always leads to the largest possible values for the weighting factors. On the other hand, there exist limits, beyond which adding more damping is no longer worthwhile. In order to optimize a realistic automotive brake, maybe not only the optimum point of the parameters but also such limit lines are existent and need to be considered to find a technically relevant solution.

Another similarity of the FE brake model compared to the minimal model can be seen in Fig. 4.6. Expanding the range of the viscous damping ($0 < \alpha_1 < 10^5$)

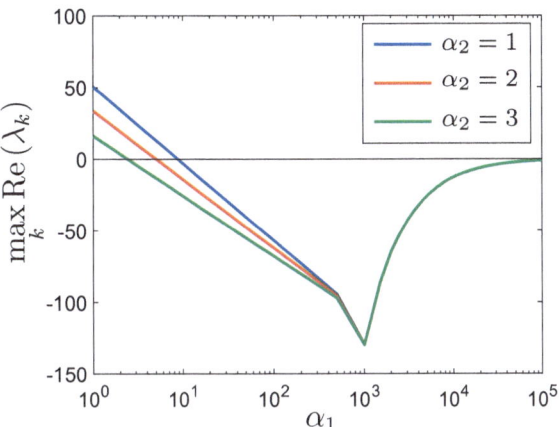

Fig. 4.6 Maximum real part over $\mathbf{D}_V$ (weighting factor α_1) for different g (weighting factor α_2); $k = 32, m = 128$, $f_{\mathrm{lim}} \approx 3\,\mathrm{kHz}$

shows that adding damping may lead to a growing maximum real part. Choosing α_2 arbitrarily, the lines in Fig. 4.6 show uniform courses having a turning point at $\alpha_1 \approx 1000$. Numerical investigations suggest that the lines converge to the zero line without passing and thus this turning point can be assumed to be a global optimum in $\alpha_1 \in \mathbb{R}^+$. Since these optimum points would lead to thousand times higher entries in the viscous damping matrix $\mathbf{D}_V$, a technical relevance is questionable. However, an analogy to the minimal model may be noticed.

4.2.2 Discussion

With the outlined optimization method time-periodic systems cannot be analyzed which is a disadvantage, as realistic automotive brakes often have cooling channels or holes resulting in time-periodic matrices in the equations of motion. However, using FLOQUET theory in connection with FE models leads to difficulties as well. PERMAS, but also all other FE codes known to the authors, offer no possibility to compute the *monodromy matrix* of a time-periodic system. Its numerical computation for large systems in a FE environment is still topic of current research.

Another limitation of the investigated FE model is that nonlinear effects, e.g. a loss of contact between the pads and the disc, nonlinear material behavior, or nonlinear properties of mechanical joints, are not taken into account. Of course, these nonlinearities can highly influence equilibrium points and thus the stability analysis [5]. But, due to time consuming and expensive computations for large FE models of automotive disc brakes, they are not considered in this study.

The third problem concerns the usage of structural damping while studying time-periodic systems. Since structural damping is defined for a single-degree-of-freedom system, Eq. (2.36) to calculate an equivalent viscous damping matrix is only an approximation, but still in use in automotive industry. For $\Omega > \Omega_{\mathrm{ref}}$ the effective damping is overestimated and vice versa. Furthermore, structural damping is based on a particular solution with a single frequency of an excitation function. Homogeneous time-periodic systems show more than one frequency and an additional problem occurs.

In this study, only those modes belonging to frequencies within a specific frequency range where brake squeal mainly occurs are optimized. However, a connection between the software packages MATLAB and PERMAS is missing. The upper frequency limit f_{lim} needs to be input manually to generate the corresponding modal reduced matrices which later can be loaded into the optimization code. An improvement would be a connection in such a way that all input values can be chosen in MATLAB.

4.3 High-Dimensional Industrial Model

In a further step, the method of optimizing the damping properties is applied to a high-dimensional FE brake model used in industry and consisting of 104 000 DOF. In contrast to the simplified model, the disc now has cooling channels to avoid overheating by strong braking manoeuvres, cf. Fig. 4.7. For the contact between the disc and the pads COULOMB's law with a constant and isotropic sliding friction coefficient $\mu = 0.8$ and no sticking is assumed; the prestress of the pads is $p = 0.8$ MPa. The model consists of the parts listed in Table 4.2, where every component has its own material properties corresponding to an individual structural damping factor g ranging from 5×10^{-3} for aluminum and steel to 10×10^{-3} for complex composite and cast iron. The structural damping matrices associated with the different components and structural damping factors are optimized individually. The upper frequency limit is set to $f_{lim} = 6$ kHz. Then, the number of eigenvalues that need to be calculated is $m = 178$ and thus $k = 70$. The problem is studied here for the case of constant coefficients.

In PERMAS, the disc's angular velocity can be varied after the matrices have been calculated using the scaling factors in Table 4.1. Still, there remains the question

Table 4.2 Components and structural damping factors of the FE brake model in Fig. 4.7 as used in industry

Component	Material	Structural damping factor g
Disc	Cast iron	$g_d = 10 \times 10^{-3}$
Pads	Complex composite	$g_p = 10 \times 10^{-3}$
Backing plate	Steel	$g_{bp} = 5 \times 10^{-3}$
Bolt	Steel	$g_b = 5 \times 10^{-3}$
Caliper	Aluminum	$g_c = 5 \times 10^{-3}$
Piston	Aluminum	$g_{pi} = 5 \times 10^{-3}$

Fig. 4.7 Industrial FE brake
model with 104 000 DOF

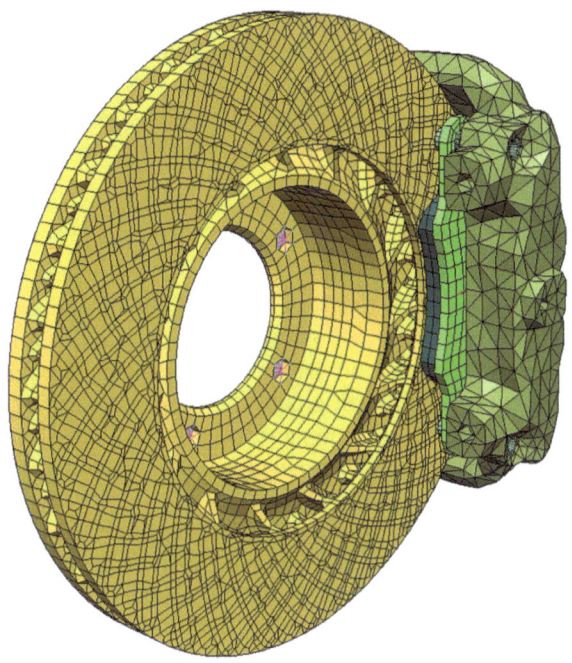

Fig. 4.8 Dependency of the
maximum real part on the
car speed

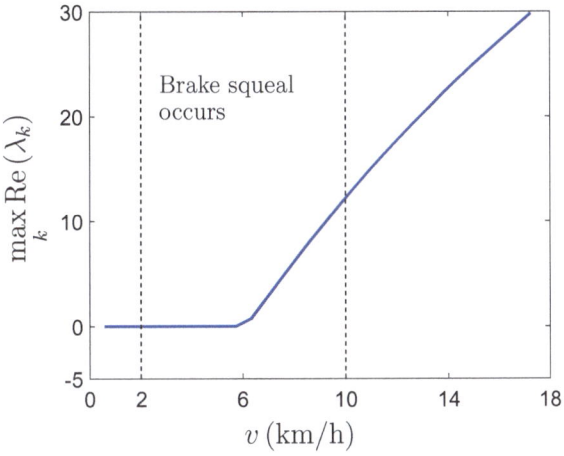

which value for ω_j should be used. Similar as in the minimal model, the friction
coefficient between the disc and the pads is assumed to be constant and thus the
approach can only be used for small angular velocities of the rotating disc. However, since brake squeal mainly occurs at low vehicle speeds (usually between 2 and
10 km/h), this assumption is no obstacle. Plotting the maximum real part over the
car speed yields the results in Fig. 4.8.

Contrary to expectations, the maximum real part grows with growing v beyond a certain velocity threshold. Within the velocity range of interest, a conservative choice then is $v = 10\,\text{km/h}$, where $\omega_j \approx 11\,\text{s}^{-1}$ assuming 20 inch wheels. In the following, the reference values are chosen as $f_{\text{cev}} = 3000\,\text{Hz}$ and $\omega_{\text{ref}} = 5\,\text{s}^{-1}$.

4.3.1   Optimization Results

Since there are in total six components to be optimized, a visualization of the optimum in a two dimensional plot is not possible. Of course, it is possible to simultaneously optimize six parameters in general. But, in order to get an impression of the influence of the different components on stability, some two dimensional results with respect to two components each are discussed in the following. In the corresponding contour plots, which are combined with a gradient field, the calculated optimum is evaluated.

At first, the optimization of the structural damping factors of the disc (weighting factor α_1) and the pads (weighting factor α_2) yields the result shown in Fig. 4.9.

The optimum point contains the largest values of the weighting factors within the admissible area. Using the values of Table 4.2, the optimized structural damping factors are $g_{d,\text{opt}} = 30 \times 10^{-3}$ and $g_{p,\text{opt}} = 30 \times 10^{-3}$. However, the corresponding maximum real part is approximately zero and hence the system is (approximately)

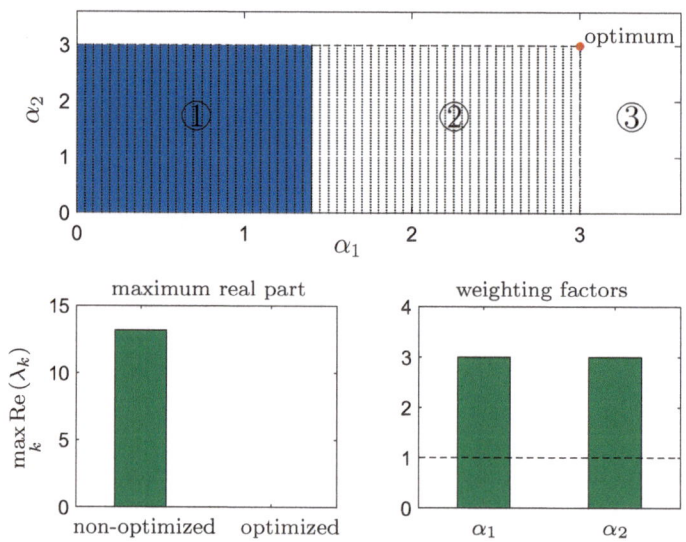

Fig. 4.9 Optimization of g_d (weighting factor α_1) and g_p (weighting factor α_2); $k = 70$, $m = 178$, $f_{\text{lim}} \approx 6\,\text{kHz}$, $v \approx 10\,\text{km/h}$, $\mathbf{\Gamma}$ and $\mathbf{c}$ are chosen as in (3.4). ①: flutter instability, admissible; ② \ ①: asymptotic stability, admissible; ③: asymptotic stability, inadmissible

Fig. 4.10 Gradient field and contour plot of g_d (weighting factor α_1) and g_p (weighting factor α_2); $k = 70$, $m = 178$, $f_{\lim} \approx 6\,\text{kHz}$, $v \approx 10\,\text{km/h}$

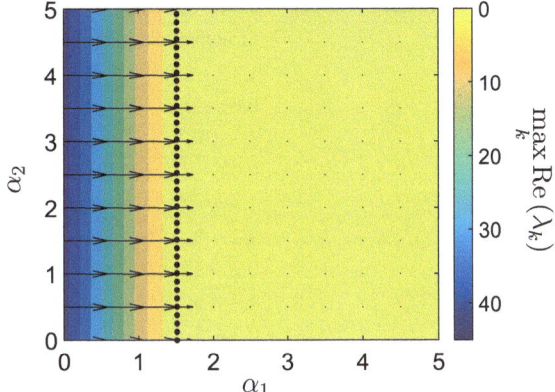

weakly stable. The question, if there are damping factors g_d or g_p large enough to make the brake model more stable, i.e. asymptotically stable, may be answered by Fig. 4.10. Similar to the simplified low-dimensional FE brake model discussed in Sect. 4.2, there is a line, above which adding damping is no longer worthwhile. The maximum real part almost stops decreasing for $\alpha_1 > 1.5$, i.e. $g_d > 15 \times 10^{-3}$. With regard to Fig. 4.5, the dotted line again represents an estimate by the authors, above which more damping may be inefficient. Of course, using the calculated optimum point leads to the minimum maximum real part in the admissible area. However, the improvement of setting $\alpha_1 = 3$ instead of $\alpha_1 = 1.5$ is of magnitude $\mathcal{O}(10^{-3})$ and hence, choosing the larger value for α_1 may be the wrong decision with respect to a cost-benefit analysis. Moreover, the contour lines are parallel to the α_2-axis so that adding more damping in the pads is also inefficient and $\alpha_1 \approx 1.5$ and $\alpha_2 = 1$ (unchanged) may be interpreted as optimum weighting factors for the disc and the pads.

In Table 4.3, the structural damping factors of aluminum, steel, and cast iron found in the literature are listed. Depending on the reference, the values for the respective material vary over several magnitudes. This makes it difficult to evaluate the technical relevance and feasibility of the optima calculated in this study. For example, considering the structural damping factor $g_d = 15 \times 10^{-3}$ to be the optimum for a brake disc made of cast iron, the values found in the literature give contradictory statements. On the one hand, [6] predicts this value to be in a possible range and on the other hand, [7] says the opposite. However, the exact value of g_d does not play a big role as long as it is larger than $g_d = 15 \times 10^{-3}$. Since the material damping is small, even the damping of the measurement instruments may have influence on the results and the energy dissipation in the contact surfaces may be even more important [8, 9].

Since small changes of the microstructure of the material may have a large influence on its damping properties, the structural damping factor cannot be defined as a material constant [9, 15]. Hence, in the next step, the cast iron of the industrial brake disc needs to be studied individually to verify if any value $g_d \geq 15 \times 10^{-3}$

Table 4.3 Structural damping factors of aluminum, steel, and cast iron from literature [15]

Material	Structural damping factor g	Reference
Aluminum	1.46×10^{-5}	[10]
	$0.03 - 1 \times 10^{-4}$	[9]
	1×10^{-4}	[11]
	$0.02 - 2 \times 10^{-3}$	[6]
Steel	1×10^{-4}	[11]
	$0.2 - 3 \times 10^{-4}$	[9]
	$0.2 - 1 \times 10^{-3}$	[12]
	$1 - 8 \times 10^{-3}$	[6]
	$0.2 - 2.8 \times 10^{-2}$	[13]
Cast iron	2×10^{-3}	[11]
	$0.3 - 3 \times 10^{-2}$	[6]
	$3.6 - 4 \times 10^{-2}$	[7]
	1×10^{-1}	[14]

can be technically implemented. It should be tested whether there are any unwanted effects when changing the structural damping of the disc, e.g. an extension of the braking distance of the car. However, since the majority of studies [6, 7, 14] suggested that values larger than $g_d \geq 15 \times 10^{-3}$ are possible, the technical relevance and feasibility of the results of this study seems assured.

In order to reduce the tendency of the brake to squeal, so called anti-squeal shims, installed between the brake pads and the caliper, are frequently used [16, 17]. The idea behind this application is that this simple way of adding damping with low cost could avoid brake squeal. However, the benefit of these shims is not always clearly evidenced and often discussed by automotive experts and customers. In this brake model, adding damping using an anti-squeal shim is equivalent to adding damping in the backing plate. In Fig. 4.11, a contour plot combined with a gradient field shows the influence of the structural damping factors g_{bp} and g_c of the backing plate and the caliper, respectively, on the stability behavior of the brake system.

Considering the efficiency of anti-squeal shims, it becomes apparent that in the present case they only have little influence on the system's stability. Setting $\alpha_2 = 5$, i.e. five times higher damping in the backing plate, the maximum real part can be reduced by one and the brake systems remains unstable. In the literature, the maximum value for g_{bp} is predicted to be $g_{bp} = 2.8 \times 10^{-2}$ which is approximately equivalent to $\alpha_2 = 5$, cf. Table 4.3.

Although the shims do not stabilize the brake system at a vehicle speed of approximately $10 \, \text{km/h}$, their influence on reducing the maximum real part may be large enough to stabilize at lower speed. Setting $v = 6.5 \, \text{km/h}$, i.e. $\omega_j \approx 1.13 \, \text{s}^{-1}$, leads to the results in Fig. 4.12. For $\alpha_2 = 5$ and $\alpha_1 = 1$ the maximum real part approximately is equal to zero, whereas setting $\alpha_2 = 1$ (unchanged) and $\alpha_1 = 1$ yields $\max_k \text{Re}(\lambda_k) \approx 2.5$. Hence, the usability of anti-squeal shims could be interpreted

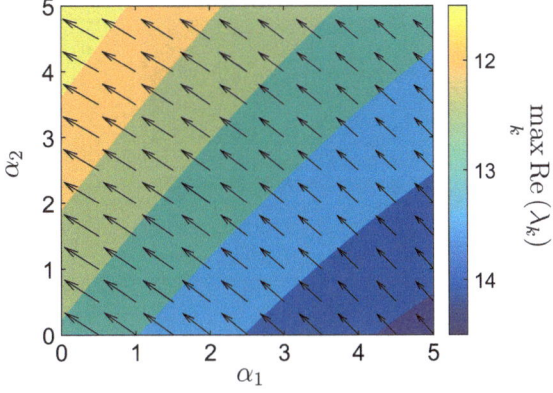

Fig. 4.11 Gradient field and contour plot of g_c (weighting factor α_1) and g_{bp} (weighting factor α_2); $k = 70$, $m = 178$, $f_{\lim} \approx 6\,\text{kHz}$, $v \approx 10\,\text{km/h}$

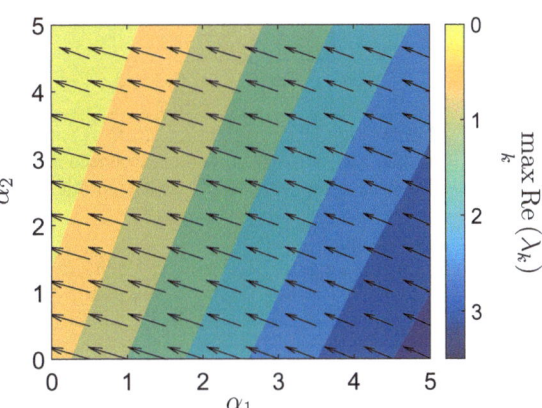

Fig. 4.12 Gradient field and contour plot of g_c (weighting factor α_1) and g_{bp} (weighting factor α_2); $k = 70$, $m = 178$, $f_{\lim} \approx 6\,\text{kHz}$, $v \approx 6.5\,\text{km/h}$

in a way that the velocity range where brake squeal occurs is delimited. On the one hand, changing the material of the shims could be more efficient. On the other hand, independent of the material, structural damping factors of magnitude $\mathcal{O}(10^{-2})$ are large with regard to Table 4.3.

As already mentioned with respect to the minimal model and the low-dimensional FE model, adding damping may also lead to a growing maximum real part. While the technical relevance of the corresponding weighting factors of these models is questionable, the destabilization paradox may occur in an interval closer to a technically relevant range in the industrial brake model. Looking at the direction of the arrows in Fig. 4.11 and Fig. 4.12, respectively, it can be seen that adding damping in the caliper (weighting factor α_1) acts destabilizing within the whole admissible range ($0 \leq \alpha_j \leq 5$).

A similar effect occurs when optimizing the structural damping factors g_d and g_c of the disc and the caliper, respectively. As demonstrated by the black arrow in Fig. 4.13, there is a region, where adding damping in the caliper may lead to flutter-instability. However, the range of values for the weighting factors is small and a

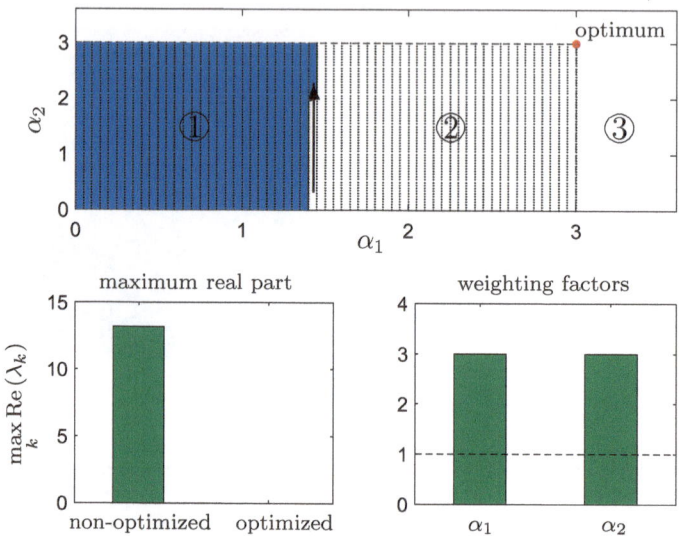

Fig. 4.13 Optimization of g_d (weighting factor α_1) and g_c (weighting factor α_2); $k = 70$, $m = 178$, $f_{\lim} \approx 6\,\text{kHz}$, $v \approx 10\,\text{km/h}$, Γ and $\mathbf{c}$ are chosen as in (3.4). ①: flutter instability, admissible; ② \ ①: asymptotic stability, admissible; ③: asymptotic stability, inadmissible

Fig. 4.14 Gradient field and contour plot of g_b (weighting factor α_1) and g_{pi} (weighting factor α_2); $k = 70$, $m = 178$, $f_{\lim} \approx 6\,\text{kHz}$, $v \approx 10\,\text{km/h}$

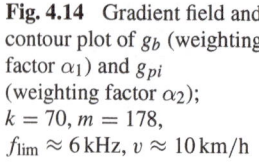

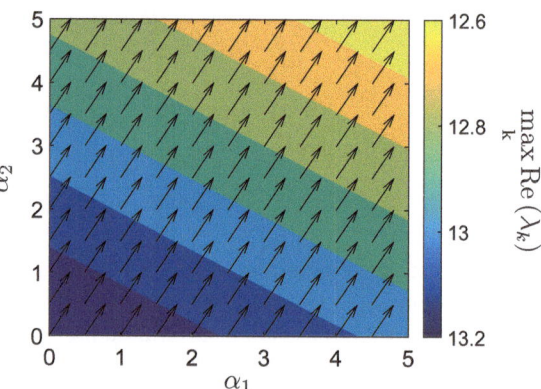

quantitative transfer to reality is questionable. Furthermore, the optimum point contains the largest possible weighting factors. Similar as in Fig. 4.9, structural damping in the disc is the dominant parameter. At the optimum point, the brake system is approximately weakly stable and adding damping in the disc using any value $\alpha_1 \geq 1.5$, independent of α_2, is stabilizing.

For the sake of completeness, the influence of structural damping in the bolt (weighting factor α_1) and in the piston (weighting factor α_2) is shown in Fig. 4.14. It is easy to see that these components are not as dominant with regard to the system's stability.

4.3.2 Discussion

To sum up the results of the optimization of the high-dimensional FE brake model from industry, the disc can be identified to be the dominant component to efficiently add damping respectively modify the structural damping factor. If $g_d \approx 15 \times 10^{-3}$, independent of the structural damping factor of any other component, the brake system is approximately weakly stable. Furthermore, it seems to be impossible to find any structural damping factor g_d large enough to make the system significantly asymptotically stable. In order to optimize the already existing brake for which the structural damping factors from Table 4.2 are assumed, the most favorable weighting factor is $\alpha_{1,opt} = 1.5$. However, measuring the structural damping factor is a technical challenge. An even more complicated task is setting it to a required value. As the differences in Table 4.3 show, evaluating the technical feasibility using values from literature or reference works is not purposeful. They can only be used as estimates. Since the microstructure of a material may have a large influence on the structural damping factor, specifically the cast iron of the brake disc needs to be studied to evaluate the technical feasibility.

Since the exact physics of structural damping is complex and not well known, it is difficult to determine a general constraint based on an economic cost function. On the one hand, Fig. 4.10 shows that adding more damping than $g_d \approx 15 \times 10^{-3}$ is not worthwhile. On the other hand, since small modifications of the material can potentially lead to much higher structural damping factors, setting for example $g_d = 40 \times 10^{-3}$ could be simpler and cheaper than setting $g_d = 15 \times 10^{-3}$. This inaccuracy of the structural damping factor could also be an explanation of the effect, that, although two brakes may seem identical, only one of them may have the tendency to squeal.

In conclusion, two aspects can be highlighted with respect to the problems and limits of optimizing the realistic FE brake model. Firstly, measuring and setting the calculated optimum points is a technical challenge. Secondly, since a FE model has thousands or even millions of DOF, modeling the structural damping, which is based on a one DOF KELVIN-VOIGT model can only be considered as a crude approximation. In addition, it is recalled that time-periodicity of the equations of motion is not taken into account here.

References

1. Ingenieurgesellschaft für technische Software. PERMAS—Workshop: Dynamik II: Erweiterungen und spezielle Anwendungen. Stuttgart (2015)
2. Ingenieurgesellschaft für technische Software. PERMAS: User's Reference Manual. Stuttgart (2014)
3. Natke, H.G.: Einführung in Theorie und Praxis der Zeitreihen- und Modalanalyse: Identifikation schwingungsfähiger elastomechanischer Systeme. Grundlagen und Fortschritte der Ingenieurwissenschaften, 3rd edn. Vieweg, Braunschweig (1992)
4. Ingenieurgesellschaft für technische Software. PERMAS: Examples Manual. Stuttgart (2014)

5. Hochlenert, D.: Nonlinear stability analysis of a disk brake model. Nonlinear Dyn. **58**(1–2), 63–73 (2009)
6. Beards, C.F.: Structural Vibration: Analysis and Damping. Arnold, London, Sidney, Auckland (1996)
7. Grote, K.-H., Feldhusen, J. (eds.): Dubbel: Taschenbuch für den Maschinenbau, 24th edn. Springer Vieweg, Berlin, Heidelberg (2014)
8. Kruse, S., Tiedemann, M., Zeumer, B., Reuss, P., Hetzler, H., Hoffmann, N.: The influence of joints on friction induced vibration in brake squeal. J. Sound Vib. **340**, 239–252 (2015)
9. Möser, M., Kropp, W., Cremer, L.: Körperschall: Physikalische Grundlagen und technische Anwendungen, 3rd edn. Springer, Berlin, Heidelberg (2010)
10. Bickel, E.: Die Metallischen Werkstoffe des Maschinenbaues, 3rd edn. Springer, Berlin, Heidelberg (1961)
11. Kollmann, F.G., Schösser, T.F., Angert, R.: Praktische Maschinenakustik. VDI. Springer, Berlin, Heidelberg, New York (2006)
12. Ungar, E.E., Zapfe, J.A.: Structural damping. In: Vér, I.L. (ed.) Noise and Vibration Control Engineering, pp. 579–609. Wiley, Hoboken (2006)
13. Müller, G., Groth C.: FEM für Praktiker—Band 1: Grundlagen: Basiswissen und Arbeitsbeispiele zur Finite-Element-Methode mit dem FE-Programm ANSYS Rev. 5.5, vol. 23 of Edition expertsoft, 7 edn. expert, Renningen (2002)
14. Lazan, B.J.: Damping of Materials and Members in Structural Mechanics. Pergamon Press, Oxford, New York (1968)
15. Niehues, K.K.: Identifikation linearer Dämpfungsmodelle für Werkzeugmaschinenstrukturen. Forschungsberichte IWB, vol. 318. Utz Herbert, München (2016)
16. Festjens, H., Gaël, C., Franck, R., Jean-Luc, D., Remy, L.: Effectiveness of multilayer viscoelastic insulators to prevent occurrences of brake squeal: a numerical study. Appl. Acoust. **73**(11), 1121–1128 (2012)
17. Kang, J.: Finite element modelling for the investigation of in-plane modes and damping shims in disc brake squeal. J. Sound Vib. **331**(9), 2190–2202 (2012)

Chapter 5
Conclusion

In this study, the main focus lies on the modification and optimization of the damping matrix of linearized disc brake models in order to stabilize or make more stable the equilibrium solution subject to sensible constraints and thus reduce the tendency of friction-induced squeal noise. Three different models are investigated numerically. Firstly, a minimal model with two degrees of freedom is analyzed to understand the qualitative stability behavior when modifying damping matrices which have a distinct physical origin, e.g. viscous, friction-induced, or material damping. In the time-invariant case, the optimization technique is based on minimizing the maximum real part of the eigenvalues, while in the time-periodic case, FLOQUET theory is used to calculate and optimize the eigenvalues (*Floquet multipliers*) of the *monodromy matrix*. The results show that the optimum is invariably located at an edge or a corner of the linear polyhedral admissible area. In a technically relevant range of parameter values, adding more damping always seems to be beneficial to make this two-degree-of-freedom minimal model more stable. However, in realistic automotive brakes, due to ventilation channels in the disc, the equations of motion are often time-periodic and the brake system may exhibit parametrically excited vibrations. With a simple numerical example it is demonstrated that complex eigenvalue analysis, which is the approach mostly used in industry, may provide wrong results when applying it to systems depending explicitly on time.

Secondly, the optimization is carried out using finite element models of simplified and realistic disc brakes whose dimension is reduced with modal truncation such that only those eigenmodes with frequencies being in the audible range are considered. In the low-dimensional system (4600 DOF), the optimum is the point of maximum damping within the admissible (rectangular) area. At a specific damping value, the maximum real part nearly stops decreasing and adding more (structural or viscous) damping is no longer worthwhile. This effect is also observed in the minimal model, where an increase of damping in the pads leads to a growing maximum real part after a certain damping threshold is exceeded. However, in the minimal model, only

© The Authors 2018
J.-H. Wehner et al., *Damping Optimization in Simplified and Realistic Disc Brakes*, SpringerBriefs in Applied Sciences and Technology, DOI 10.1007/978-3-319-62713-7_5

viscous and friction-induced damping is present so that a direct comparison is not possible.

Thirdly, the introduced optimization method is applied to a high-dimensional, industrial disc brake model (104 000 DOF), where the main focus is to optimize the structural damping factors of the different brake components. Although the optimum points are again characterized by the maximum values of damping within the admissible area, these optima could be the wrong decision with respect to a cost-benefit analysis. As observed in the minimal model and the simple finite element model, there are limits above which adding more damping does not lead to a significant improvement of the stability behavior of the brake system. Furthermore, the brake disc is the dominant component to efficiently add damping respectively modify its structural damping factor. Independent of the structural damping factors of any other component, sufficiently large structural damping in the disc seems to always stabilize the investigated brake system.